"There are two things that delight me about this book. Firstly, the knowledge, experience, and passion of those who created it. Secondly, the support it gives to educators."

— **Wendy Goucher, MSc (Res) FBCS,** *Cyber Consultant and Author, Scotland, UK*

"To foster our students' success in a digital era, it's important that we exemplify proper usage ourselves and give them chances to apply character education principles in digital spaces, which now encompass the dynamic landscape of AI. *From Street-smart to Web-wise®: A Safety Training Manual Built for Teachers and Designed for Children* serves as a resource, offering educators a pathway to integrate these two spheres and facilitate meaningful learning experiences."

— **Kristina Lodes,** *Digital Learning Coach, Saint Louis, Missouri USA*

"Online safety is one of the top concerns for educators and parents."

— **Christine Burke,** *Elementary School Principal, St. Louis, Missouri USA*

From Street-smart to Web-wise®

Book 2 continues as the tiny fingers in Book 1 Grades K-2 grow and become more familiar with online activities. The critical job of ensuring our children's safety expands as students become more independent and begin to have greater online autonomy. *From Street-smart to Web-wise®: A Cyber Safety Training Manual Built for Teachers and Designed for Children* isn't just another book — it's a passionate call to action for teachers, a roadmap to navigate the digital landscape safely, with confidence and care.

Written by authors who are recognized experts in their respective fields, this accessible manual is a timely resource for educators. Dive into engaging content that illuminates the importance of cyber safety, not only in our classrooms but extending into the global community.

Each chapter is filled with practical examples, stimulating discussion points, and ready-to-use lesson plans tailored for students in third and fourth grades. Regardless of your technology skill level, this book will provide you with the guidance and the tools you need to make student cyber-safety awareness practical, fun, and impactful.

As parents partner with educators to create cyber-secure spaces, this book stands as a framework of commitment to that partnership. It's a testament to taking proactive steps in equipping our young learners with the awareness and skills they need to tread the digital world securely.

By choosing *From Street-smart to Web-wise®: A Cyber Safety Training Manual Built for Teachers and Designed for Children*, you position yourself at the forefront of educational guardianship, championing a future where our children can explore, learn, and grow online without fear. Join us on this journey to empower the next generation — one click at a time!

From Street-smart to Web-wise®

A Cyber Safety Training Program Built for Teachers and Designed for Children (Book 2)

Al Marcella, Brian Moore, and Madeline Parisi

CRC Press
Taylor & Francis Group
Boca Raton London New York

CRC Press is an imprint of the
Taylor & Francis Group, an **informa** business

Designed cover image: © Shutterstock

First edition published 2025
by CRC Press
2385 NW Executive Center Drive, Suite 320, Boca Raton FL 33431

and by CRC Press
4 Park Square, Milton Park, Abingdon, Oxon, OX14 4RN

CRC Press is an imprint of Taylor & Francis Group, LLC

ISBN: 9781032731759 (hbk)
ISBN: 9781032738604 (pbk)
ISBN: 9781003466338 (ebk)

DOI: 10.1201/9781003466338

Typeset in Caslon
by codeMantra

To our families and friends,

for their encouragement, support, and love.

and

To the dedicated educators globally, who nurture,

guide, support, and ignite a passion for knowledge

and learning in children of every educational background.

Contents

List of Figures

Foreword

Children today have an experience of the world that is very different from that which their parents and teachers had at a young age. Technology isn't just a toy or a television, it is a very real part of the way they experience the world.

During the period of COVID, even where people were able to be mobile locally, many were still restricted to travel beyond that, especially overseas. For many young learners during this time they may have met and interacted with members of their family and friends entirely through the virtual medium and for quite a while.

The implication is that for learners today the computer is more than just a communication medium, it is a portal to the world outside of their door, and one where the door is not locked, and the handle is low enough to reach. This adds pressure on parents and educators to help children learn to keep safe and aware in the same way as they do walking down the street. If you think teaching a child to cross the road safely is hard, when they can see the risk in the form of a massive rumbling truck, then how can they understand hidden risks? We can't just talk about the witch and the apple in Snow White anymore, it takes knowledge and skill to provide educators with approaches and materials to allow them to plant the seeds of cyber safety from the youngest age and nurture it as the learner grows.

There are two things that delight me about this book. Firstly, the knowledge, experience, and passion of those who created it. The authors are recognised experts in their fields, and continue to develop by research and practice, and their attention to detail in all aspects must have been exhausting at times.

Secondly, the support it gives to educators. It is no secret that educators today are faced with a huge burden of work to try and develop their learners and kindle their understanding of the world and their desire to play their part in it with the pressure of changes in behaviours and understanding of even the youngest learners. With the constant changes that developing technologies and behaviour issues bring, an exercise that worked for a class last year may not work this year, it all must be new and fresh every day.

Thankfully this book is not a book of exercises that can become obsolete. It asks questions and in doing that it provides building bricks and tools from which educators can craft their approach for each class at each moment. It is not a crutch for educators, it is a strong ladder that will support their work with learners. I believe there is no greater aim for any work in this area than to do that. I applaud their work.

Wendy Goucher MSc (Res) FBCS
Cyber Consultant and Author
Scotland, UK

Preface

The authors developed this text for teachers and student leaders as the reader and user, and not the student who is the ultimate beneficiary of the material.

This teaching manual is designed to help make an educator's job and knowledge transfer easier by providing context to young learners and by providing relevant lesson plans for each chapter.

The approach is structured to emphasize the following:

- The creation of content in lesson plan format.
- The emphasis on skills training and associated assessments.
- The development of critical thinking skills regarding cyber hygiene which begins at the earliest ages.
- The design and development of instructional materials that can be delivered regardless of the instructor or facilitator's experience level.

In a school classroom environment, it's often best to use inclusive and neutral language when referring to children and educators, as this promotes a respectful and supportive atmosphere.

The reader will notice throughout this book that instead of using terms such as "girl, boy, he, she, boys, or girls" or using gender-specific terms, the authors will refer to children when discussing pedagogy as "students," "learners," or "young learners." These learners may be involved in any academic learning program, within any positive learning environment serving students.

When referring to teachers the authors recognize that this descriptor is intentionally broad. The authors use the term teacher or educator to collectively include instructress or instructor, lecturer, tutor, facilitator, mentor, counselor, educationalist, and trainer — i.e. a person who regularly works with children.

Examples Are Not Endorsements

This document contains references to materials that are provided for the reader's convenience. The inclusion of these references is not intended to reflect their importance, nor is it intended to endorse any views expressed, or products or services offered by third-party providers. These reference materials may contain the views and recommendations of various subject matter experts as well as hypertext links, contact addresses, and websites to information created and maintained by other public and private organizations.

The opinions expressed in any of these third-party materials included as references or examples do not necessarily reflect the positions or policies of the authors.

The authors do not control or guarantee the accuracy, relevance, timeliness, update, or completeness of any outside, third-party information included in this publication. These references should not be construed as an endorsement by the authors or by the publisher. The reader should validate and substantiate any information directly from third-party providers before authorizing the acquisition, implementation, or use of any product or service referenced in this book.

Acknowledgments

The authors wish to thank the many contributors who have provided input to this program as it transformed from a concept with various delivery methods to this resource for teachers and those persons who regularly work with children, we thank you.

The authors wish to recognize and thank the following for their contributions and for so graciously responding to requests for further information.

John F. Boyce, Head-People Development, AMSOIL Inc. Duluth, Minnesota and Adjunct Faculty, Northwestern University, Evanston, Illinois, for providing insight into book development at the earliest stages of transferring the concept to a teacher's manual.

Christine Burke, Elementary School Principal, St. Louis, Missouri, for providing feedback to CRC Press which provided insight to the authors.

The authors also thank the anonymous reviewers identified by the Publisher, CRC Press, who initially reviewed our proposed book series and provided positive feedback and comments on the proposed content when the authors initially brought the idea for this book to CRC Press.

Authors

Dr. Al Marcella, Ph.D., CISA, CISM, President of Business Automation Consultants (BAC) LLC, is an internationally recognized public speaker, researcher, IT consultant, and workshop and seminar leader with 46 years of experience in IT audit, risk management, IT security, and assessing internal controls, having authored numerous articles and 30 books on various IT, audit, and security-related subjects. Dr. Marcella's clients include organizations in financial services, IT, banking, petrol-chemical, transportation, services industry, public utilities, telecommunications, and departments of government and nonprofits. Dr. Marcella is also a tenured, full-time professor at Webster University, teaching at the Walker School of Business.

Research conducted by Dr. Marcella on unmanned aircraft systems, cyber extortion, workplace violence, personal privacy, electronic stored information, privacy risk, cyber forensics, disaster and incident management planning, the Internet of Things, ethics, and astrophotography has been published in the *ISACA Journal*, *Disaster Recovery Journal*, *Journal of Forensic & Investigative Accounting*, *EDPACS*, *ISSA Journal*, *Continuity Insights*, *Internal Auditor Magazine* and the Astronomical League's *Reflector Magazine*. Dr. Marcella's fourth book on cyber forensics, *Cyber Forensics: Examining Emerging and*

Hybrid Technologies, a collaborative effort written with a team of six co-authors, was published by CRC Press in 2021.

Dr. Marcella holds a BS degree in Management, a BS degree in Information Technology Management, an MBA with a concentration in Finance, and a PhD in Management/Information Technology Management. Dr. Marcella is a Certified Information Systems Auditor (CISA) and a Certified Information Security Manager (CISM) and holds an ISACA Cybersecurity Certificate.

Dr. Marcella is the 2016 recipient of the Information Systems Security Association's Security Professional of the Year award and recipient of the Institute of Internal Auditors Leon R. Radde Educator of the Year 2000 award and has been recognized by the Institute of Internal Auditors as a distinguished adjunct faculty member. Dr. Marcella has taught IT audit seminar courses for the Institute of Internal Auditors (IIA) and the Information Systems Audit and Control Association (ISACA).

Brian Moore is a passionate certified K-8 general education teacher with over 20 years of experience developing and implementing diverse curriculums covering wide range of subjects. Moore is highly skilled at motivating students through positive encouragement and reinforcement of concepts via interactive classroom instruction and observation. Moore is successful in helping students develop strong literacy, numeracy, social, and learning skills.

Moore has valuable experience in classroom administration, professional development, and project planning in one of Phoenix Arizona's largest Unified School Districts. Prior to the classroom, Moore was the site director for the Scottsdale/Paradise Valley (AZ) YMCA's Before and After School Program.

Equipped with extensive background in versatile education environments, a student-centric instructor, academic facilitator, and motivational coach, Moore is well-versed in classroom and online technologies.

Moore received both his Bachelor of Arts and Master of Education: Educational Leadership from Northern Arizona University — Flagstaff, Arizona, and holds Certifications: Standard Professional Elementary Education, K-8 and Pre-K-12 Principal, and the Endorsement: Structured English Instruction, K-12.

Madeline Parisi, M.Ad.Ed., Principal, founded Madeline Parisi & Associates LLC (MPA) in 2013, an international organization providing business training content and training material, after a distinguished career working with professional associations. Together with a pool of subject matter experts, MPA provides business training materials, in-house and virtual training, white-label writing services, and professional certification training and certification exam question development services.

Parisi is a recognized adult education content developer with a 30-plus year career in business training and professional development serving finance, legal, and IT audit, IT security, and risk management professionals.

Parisi holds a Bachelor of Arts degree in Criminology from the University of Illinois, Chicago and a Master of Adult Education from National Louis University, Wheeling, Illinois. Parisi also holds certificates in Volunteer Management from Harper College, Palatine, Illinois and additional certificates in Organizational Development, Lean Six Sigma, and Project Management. Parisi is the author of several columns and articles published in various trade publications and along with Al Marcella is the co-author of the white paper "Assessing, Managing and Mitigating Workplace Violence: Active Shooter Threat."

Partnering with Al Marcella, The Training Resource Center, LLC, is a jointly operated entity providing training and consulting services, specializing in Environment, Social, and Governance (ESG).

1

Character Education

Character education plays a vital role in the academic setting for children in the third through fourth grades. It encompasses the development of important personal and ethical qualities that extend beyond academic achicvements. In these critical years of early education, character education serves as a foundation for building responsible, empathetic, and morally sound individuals. By nurturing values such as respect, integrity, empathy, and responsibility, educators aim to equip students with the essential life skills needed to succeed not only in school but also in their future endeavors.

Introduction

Character education creates a positive and inclusive learning environment that fosters a sense of belonging and emotional well-being among students. When children in the third and fourth grades are encouraged to understand and practice virtues like kindness and perseverance, they become not only better learners but also better citizens.

Character education helps students develop essential social skills, improve classroom behavior, and enhance their ability to collaborate and communicate effectively with peers and teachers. Moreover, it empowers young learners to make ethical decisions, resolve conflicts amicably, and navigate the challenges they encounter during their educational journey.

Ultimately, character education in the third and fourth grades contributes to the holistic development of students, molding them into well-adjusted individuals who can positively impact their communities and society at large.

DOI: 10.1201/9781003466338-1

Character Education

What Is Character?

Character can be defined in various forms, depending on personal contexts. Some examples include:

- Understanding, caring about, and acting upon core ethical values.
- The set of characteristics that motivate and enable one to function as a moral agent, do one's best work, effectively collaborate in the common space to promote the common good, and inquire about and pursue knowledge and truth.
- A set of personal virtues that produce specific moral emotions, inform motivation, and guide conduct.
- The traits and moral or ethical qualities distinctive to an individual.[1]

How Do We Define Character Education?

Character education refers to the deliberate effort to foster positive qualities and values in individuals, to help them develop good character traits. It goes beyond academic instruction and focuses on instilling virtues, ethical principles, and social skills that contribute to personal and social well-being. Character education aims to shape individuals into responsible, respectful, and compassionate members of society.

Key components of character education often include:

- Values and Virtues: Teaching and emphasizing the importance of core values such as honesty, integrity, responsibility, fairness, respect, and empathy. These virtues form the foundation of good character.
- Ethical Decision-Making: Providing individuals with the tools and skills to make ethical decisions in various situations. This involves critical thinking, problem-solving, and considering the consequences of one's actions.
- Social and Emotional Learning (SEL): Incorporating elements of social and emotional education to help individuals understand and manage their emotions, develop empathy, and build positive relationships with others.

- Civic and Global Responsibility: Encouraging a sense of responsibility toward one's community and the larger world. This includes promoting civic engagement, environmental awareness, and a commitment to social justice.
- Resilience and Perseverance: Teaching individuals how to cope with challenges, setbacks, and failures, fostering resilience and perseverance in the face of adversity.
- Positive Behavior Reinforcement: Recognizing and reinforcing positive behaviors to create a positive and supportive learning environment. This can involve praise, rewards, and other forms of positive reinforcement.
- Role Modeling: Demonstrating positive character traits through the actions and behavior of teachers, administrators, and other influential figures in a person's life. Role modeling is a powerful way to convey the importance of good character.
- Community Involvement: Engaging individuals in activities that promote community service and collaboration, fostering a sense of responsibility toward others, and building a strong sense of community.

Character education is often integrated into formal educational curricula and can be implemented in various settings, including schools, families, and community organizations. The ultimate goal is to develop individuals who not only excel academically but also contribute positively to society through their character and ethical conduct.[2]

Instilling Values through Education

The benefits of character education are manifold. Students gain life skills that help them manage relationships and conduct themselves with honor. Schools see improved academic performance and behavior when positive values are instilled. Ultimately, character education aims to develop mature citizens who contribute to a just and caring society. The values we cultivate in our youngest generations will chart the moral course for the future.

Character education actively teaches core ethical and moral values in an intentional, proactive manner. Character education furnishes a more holistic form of schooling that enables students to become healthy, balanced, civic-minded adults. Academic institutions of all types and levels should actively investigate and create caring communities and collaborations with parents to support character development.

Eleven Principles of Effective Character Education

Character education is a broad and evolving field, and there are various principles and frameworks that educators and researchers use to guide character education programs. One widely recognized framework is the "Eleven Principles of Effective Character Education" developed by the Character Education Partnership (now merged with Character.org).

These principles provide a foundation for effective character education programs and actively promote core ethical values like respect and performance values like diligence. They broadly define character beyond just morals.

Many school leaders also use the 11 Principles as a school improvement process. The 11 Principles focus on all aspects of school life, including school culture and climate, social and emotional learning (SEL), student engagement and academic achievement, as well as Multi-Tiered System of Supports (MTSS), Positive Behavioral Interventions and Supports (PBIS), Response to intervention (RTI), restorative practices, teacher morale, and parent engagement.

MTSS offers a framework for educators to engage in data-based decision-making related to program improvement, high-quality instruction and intervention, SEL, and positive behavioral supports necessary to ensure positive outcomes for districts, schools, teachers, and students.[3]

PBIS is an evidence-based, tiered framework for supporting students' behavioral, academic, social, emotional, and mental health. When implemented with fidelity, PBIS improves social-emotional competence, academic success, and school climate. It also improves teacher health and well-being. It is a way to create positive, predictable, equitable, and safe learning environments where everyone thrives.[4]

RTI aims to identify struggling students early on and give them the support they need to thrive in school.

PBIS is a specific approach to behavior management, and MTSS is a broader framework that includes academic and behavioral supports, of which PBIS is just one component.

RTI is considered a narrower approach than MTSS. An RTI approach focuses solely on academic assessments, instruction, and interventions. MTSS is a comprehensive framework that includes academic, behavioral, and social-emotional support.[5]

Eleven Principles in Schools

The Eleven Principles in Schools (11 Principles) serve as guideposts for schools to plan, implement, assess, and sustain their comprehensive character development initiative.

The 11 main principles are:

Principle 1: Core values are defined, implemented, and embedded into school culture.

Schools that effectively emphasize character development bring together all stakeholders to consider and agree on specific character strengths that will serve as the school's core values.

Principle 2: The school defines "character" comprehensively to include thinking, feeling, and doing.

The "core values" of a school serve as touchstones that guide and shape a student's thinking, feelings, and actions.

Principle 3: The school uses a comprehensive, intentional, and proactive approach to develop character.

Schools committed to character development look at all they are doing through a character lens. They weave character into every aspect of the school culture.

Principle 4: The school creates a caring community.

A school committed to character development has developed an "ethic of caring" that permeates the entire school.

Principle 5: The school provides students with opportunities for moral action.

Through meaningful experiences and reflection opportunities, schools with a culture of character help students develop their commitment to being honest and trustworthy, volunteering their time and talents to the common good, and when necessary, showing the courage to stand up for what is right.

Principle 6: The school offers a meaningful and challenging academic curriculum that respects all learners, develops their character, and helps them succeed.

Because students come to school with diverse skills, interests, backgrounds, and learning needs, an academic program that helps all students succeed will be one in which the content and pedagogy engages all learners and meets their individual needs. This means providing a curriculum that is inherently interesting and meaningful to students and teaching in a manner that respects and cares for students as individuals.

Principle 7: The school fosters students' self-motivation.

This principle emphasizes intrinsic motivation over extrinsic motivation. Character means doing the right thing and doing your best work even when no one is looking.

Principle 8: All staff share the responsibility for developing, implementing, and modeling ethical character.

All school staff share the responsibility to ensure that every young person is practicing and developing the character strengths that will enable them to flourish in school, in relationships, in the workplace, and as citizens.

Principle 9: The school's character initiative has shared leadership and long-range support for continuous improvement.

Schools of character have leaders who visibly champion the belief expressed by Martin Luther King Jr. that *Intelligence Plus Character — That is the Goal of a True Education*. These school leaders establish a Character Committee — often composed of staff, parents, community members, and students — and give the Committee the responsibility to design, plan, implement, and assess the school's comprehensive character development initiative.

Principle 10: The school engages families and the community as partners in the character initiative.

Schools of character involve families. Parents are encouraged to reinforce the school's core values at home. School leaders regularly update families about character-inspired goals and activities.

Principle 11: The school assesses its implementation of character education, its culture and climate, and the character growth of students regularly.

Schools of character use a variety of approaches to assess the character development of their students, including student behavior data and surveys. Schools also assess the culture and climate of the school, focusing particular attention on the extent to which the school's core values are being emphasized, modeled, and reinforced.[6]

Importance of Character Development in Individuals

Character education actively prepares students for modern life's challenges. It provides balance to media and Internet messages. As other influences decline, schools actively offer stable community values. Character education principles are traditional yet actively equip students for the future by providing core values with academics to develop responsible citizens.

While academically correct, how does one explain the "what" of character development in terms that grade school children will understand? Well, one does so by transferring the principles of character development into meaningful examples these students can relate to.

The concept of character education in the third and fourth grades is more easily understood by young learners when discussed in age-appropriate terms and with age-appropriate examples.

As an example, one approach to beginning this discussion may proceed as follows…

Character education is about learning important values and qualities that help you become a good person and make good choices in life. Let's take the value of "kindness" as an example. Imagine you see someone in your class who is feeling sad because they dropped their lunch and it spilled everywhere.

Being kind means you could help them pick up their things and maybe share some of your snacks with them. It's about thinking about how others feel and trying to make things better for them.

By learning about kindness and other values, you can grow up to be someone who makes the world a better place for everyone! (see Figure 1.1)

Figure 1.1   An act of kindness.[7]

So, character education is like a special set of skills and values that you carry with you every day. It's about being kind, and respectful, and making choices that make you proud. Just like a garden needs care to grow, your character needs attention and practice to become strong and wonderful!

Character development empowers young minds with a moral compass. By instilling values like honesty and responsibility, educators equip students with the tools to make decisions that reflect integrity. These foundational virtues become guiding lights, steering students away from impulsive actions and toward thoughtful, considerate behavior.

Importance of Character Development in Education

One of the great education reformers, Horace Mann, in the 1840s, helped to improve instruction in classrooms across the United States (U.S.), advocating that character development was as important as academics in American schools.

The U.S. Congress, recognizing the importance of this concept, authorized the Partnerships in Character Education Program in 1994. The No Child Left Behind Act of 2001 renews and re-emphasizes this tradition — and substantially expands support for it.

One of the six goals of the Department of Education is to "promote strong character and citizenship among our nation's youth."[8] To reach this goal, the U.S. Department of Education joins with state education agencies and school districts across the U.S. to provide vital leadership and support to implement character education.

To successfully implement character education, schools are encouraged to:

- Take a leadership role to bring the staff, parents, and students together to identify and define the elements of character they want to emphasize.
- Provide training for staff on how to integrate character education into the life and culture of the school.
- Form a vital partnership with parents and the community so that students hear a consistent message about character traits essential for success in school and life.
- Provide opportunities for school leaders, teachers, parents, and community partners to model exemplary character traits and social behaviors.[9]

Educators play a crucial, active role in character building by providing guidance and serving as role models, although parents are the most influential. Schools actively give students opportunities to learn values through peer interactions.

Importance of Character Development for Parents and Communities

Character education actively supports parents' influence rather than supplanting it. Educators and parents should be active partners in instilling values and providing stability amid other influences. Both parents and students actively benefit from character education. It provides students with a moral compass amongst competing messages. Many parents actively want values-based education.

With work and frequent moves disrupting community ties, schools are actively one of the few stable influences. Character education makes schools actively values-based stable influences. Character education's holistic development of academics, values, and character is critically essential for the next generation. It actively complements parents' role comprehensively.

Character building in adolescents is insignificantly influenced by parenting. Parents, through parenting, will shape the character of the child. In line with their development and age, children become teen-agers and will expand their socialization. As a result, their psycho-social life is also developed. This happens because the scope of their association influences the psychosocial development of adolescents.[10]

The Foundation of Character — Core Values

In the pursuit of building a foundation of character, a set of 14 core values (see Figure 1.2) emerges as the guiding principles that shape individuals into compassionate and responsible members of society.

At the heart of this philosophy lies the value of Caring, promoting an innate sense of concern for the well-being of others. Citizenship, an acknowledgment of one's role in a broader community, intertwines

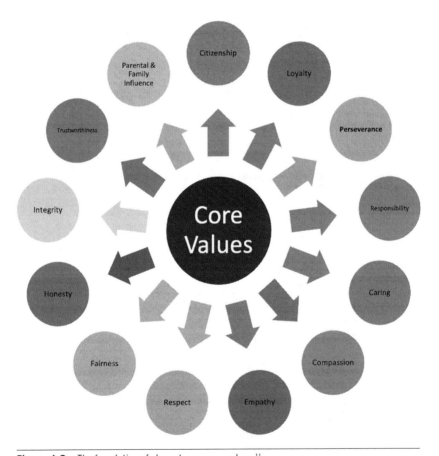

Figure 1.2   The foundation of character — core values.[11]

seamlessly with the understanding of The Role of Family. Recognizing the profound impact of upbringing and Parental Influence on character development. Compassion and Empathy, cornerstones of human connection, emphasize the importance of understanding and sharing in the experiences of others, fostering a sense of unity and interconnectedness.

Fairness, Honesty, and Integrity form the ethical triad, emphasizing the importance of moral rectitude in all aspects of life. Loyalty, an enduring commitment to relationships and shared values, joins individuals together in a bond of trust and mutual support. Parental Influence, as a separate entity, underscores the pivotal role that family plays in shaping character, serving as a backpack of values and principles that individuals carry into the world. Perseverance, an unwavering commitment to overcoming challenges, stands as a testament to the strength and resilience inherent in an individual's character.

Respect, an acknowledgment of the intrinsic worth of each person, creates an environment of dignity and understanding. Responsibility, both to oneself and to others, manifests as a commitment to contribute positively to society. Trustworthiness, the bedrock of any meaningful relationship, underscores the importance of reliability and honesty in fostering enduring connections.

Together, these 14 core values coalesce to form a solid foundation upon which individuals can build a character marked by empathy, integrity, and a commitment to the well-being of the larger community.[12]

A Student's Character Education Credo

- Character means doing the right thing. When I am at school or home, I should be honest, kind, fair, and responsible. This helps me be a good friend and classmate. My teachers say developing character is very important.
- Being honest means telling the truth. I will not lie or cheat on schoolwork. Telling the truth even when I make a mistake shows I have integrity. Honesty makes people trust me.
- I also want to be kind. I treat others nicely and help my friends when they need it. Being kind makes my classmates and me feel happy. We get along better when we are kind.

- It's important to be fair too. I take turns and share with others. When there is a problem, I listen first before reacting. Being fair means caring about how others feel.
- Finally, I want to be responsible. That means doing my part, cleaning up, and taking care of my belongings. Responsibility gives me confidence. My family and teachers are proud when I am responsible. Good character helps me become my best self.

Ethics: A General Introduction and Overview

Before we provide a generally accepted definition of ethics, we must examine the core principles that when working together, lead to this generally accepted definition of ethics.

Core principle #1 — Values. Values are the foundation of an individual person's ability to judge between right and wrong, they frame the decisions we make. Values include a deep-rooted system of beliefs that guide a person's decisions. They form a personal, individual foundation that influences a particular person's behavior.[13]

Core principle #2 — Morals. Morals are the principles that guide individual conduct within society, and they emerge out of core values. While morals may change over time, they remain the standards of behavior that we use to judge right and wrong.

In the study of ethics, the terms amoral and immoral are often interchanged; however, in application, they are quite different.

Amoral refers to a lack of moral principles or a disregard for moral values altogether. When someone or something is described as amoral, it means they do not differentiate between right and wrong, nor do they consider ethical implications in their actions.

Immoral, on the other hand, refers to behavior that is in direct violation of accepted moral principles or standards. Immoral actions are considered wrong or unethical and often lead to negative consequences for others or society as a whole.

Amoral behavior lacks a moral compass altogether, whereas immoral behavior knowingly goes against established moral standards. While amoral actions may not carry inherent moral judgment, immoral actions are generally seen as wrong and ethically objectionable.

Core principle #3 — Judgment. Judgment refers to the process of evaluating and making decisions about the moral rightness or wrongness of an action, behavior, or belief. Ethical judgment involves considering various moral principles, values, and standards to assess whether a particular action is morally permissible, obligatory, or prohibited. Ethical judgment consists not of getting the right answers all the time, but…. of consistently asking the right questions.

Core principle #4 — Norms. Norms refer to the established standards, rules, or principles that govern and guide human behavior within a particular social or cultural context. These norms dictate what is considered morally acceptable or unacceptable within a given community, group, or society. Norms serve as a framework for evaluating and judging the ethicality of actions, decisions, and practices. Norms help to promote ethical conduct and help prevent harmful or unethical behaviors.[14]

Core principle #5 — Standards. Standards refer to the specific criteria or benchmarks that are used to assess and measure ethical behavior or the moral quality of actions, decisions, or practices. These standards serve as guidelines for evaluating whether an individual, organization, or society is adhering to ethical principles and behaving in a morally responsible manner. Standards are crucial in promoting ethical behavior and holding individuals and institutions accountable for their actions, ensuring that they align with recognized ethical values and principles (see Figure 1.3).

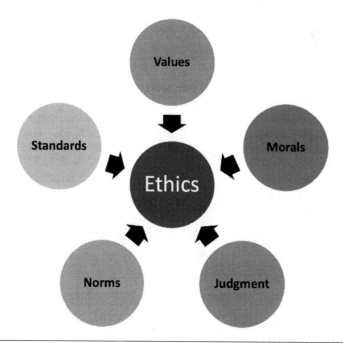

Figure 1.3   Core principles of ethics.[15]

Values are "judgments of worth," moral principles, which should have a certain weight in the choice of an action. Morals refers to what is judged as right, just, or good. Judgment is the ability to make accurate determinations. Norms state what is morally correct behavior in a certain situation. Standards assess and measure ethical behavior. Working together these five core principles lead us to a generally accepted definition of ethics.

Ethics ... the collection of values, morals, judgments, norms, and standards, which provides a framework for acting.

In their article "A Framework for Ethical Decision Making," the team from Markkula Center for Applied Ethics at Santa Clara University states that "ethics refers to standards and practices that tell us how human beings ought to act in the many situations in which they find themselves—as friends, parents, children, citizens, businesspeople, professionals, and so on. Ethics is also concerned with our character. It requires knowledge, skills, and habits."

The team goes on further in defining ethics by identifying what ethics is not. One characteristic of what ethics is not, and which is receiving much attention and debate, concerning GEN-AI models is... ethics is not science. "Social and natural science can provide important data to help us make better and more informed ethical choices. But science alone does not tell us what we ought to do. Some things may be scientifically or technologically possible and yet unethical to develop and deploy."[16]

Final Thoughts on Character Education and Cyber Safety

In today's digital age, where technology and the Internet play a significant role in education, fostering good character traits becomes crucial to ensure that young learners develop responsible and safe online behaviors.

One of the key teaching principles at the intersection of character education cyber safety is teaching children about digital citizenship. This includes instilling values such as respect, empathy, and responsible behavior when using online platforms and interacting with others digitally. Through character education, students can learn the importance of treating their online peers with kindness and respect, just as they would in face-to-face interactions.

By understanding the impact of their words and actions online, they can become more responsible digital citizens, which is essential for their safety and the safety of others in the virtual world.

Character education helps children in the third and fourth grades develop critical thinking skills and the ability to make informed decisions online. They learn to recognize potential risks and understand the consequences of their actions, which is vital in navigating the Internet safely.

By promoting values like honesty and integrity, character education encourages students to be honest about their online activities, report any harmful behaviors they encounter, and seek help when needed. Overall, character education in the context of cyber safety equips young learners with the moral compass and ethical foundation to engage with technology responsibly, protecting themselves and fostering a positive online environment for all.

Behavior — What Is Acceptable/Unacceptable?

Anti-Bullying

What Is Bullying?　The Anti-Bullying Alliance and its members have an agreed shared definition of bullying based on research from across the world over the last 30 years.

> The repetitive, intentional hurting of one person or group by another person or group, where the relationship involves an imbalance of power. Bullying can be physical, verbal, or psychological. It can happen face-to-face or online.[17]

School bullying takes on many forms and may take place in various locations; however, the initial relationship between parties was formed in a school setting. Bullying may also be in the form of direct, in-person contact whereas it is more verbal or indirect contact such as cyberbullying. It is important to ensure that students understand that teachers, counselors, and all support systems within the school, or agency, are available for help and counseling regardless of where the bullying activity takes place, making school a safe environment.

Bullying impacts not only the ability to focus and learn but also physical and mental health, both while the bullying activity is taking place and in the long term.

The monitoring of bullying activities must be an ongoing activity in the classroom, on school grounds, and during school-sanctioned activities, and is supported school-wide. Successful monitoring requires a unified classroom plan that is aligned with and supported by the school's administrative plan.

The Centers for Disease Control and Prevention (CDC) categorizes bullying as a form of youth violence and an adverse childhood experience (ACE). The CDC further indicates that bullying is a frequent discipline problem with nearly 14% of public schools reporting that bullying is a discipline problem occurring daily or at least once a week and cites the following statistics.

- Reports of bullying are highest in middle schools (28%) followed by high schools (16%), combined schools (12%), and primary schools (9%).
- Reports of cyberbullying are highest in middle schools (33%) followed by high schools (30%), combined schools (20%), and primary schools (5%).[18]

So prevalent, all 50 U.S. and many U.S. Commonwealths and Territories have anti-bullying laws and/or policies.[19]

The following discussion and the lesson plans included in this chapter incorporate the following anti-bullying types and definitions. This is not an all-inclusive list of anti-bullying types and is a compilation of bullying types from the University of the People[20] and the Preventing and Promoting Relationships & Eliminating Violence Network (PREVNet).[21]

Although many of the types of bullying listed may appear to be direct, in-person bullying, they may also take place indirectly online.

Cyberbullying Cyberbullying is any type of bullying that happens online. It can be hurtful comments on a personal site or deceptive private messaging. Includes the use of email, cell phones, text messages, and Internet sites to threaten, harass, embarrass, socially exclude, or damage reputations and friendships.

Physical Bullying Physical bullying always involves physical contact with the other person. This can mean hand-to-hand, but it can also mean throwing items, tripping, or eliciting others to cause physical harm to a person. This type of bullying can also cause harm to property and belongings.

Verbal Bullying Verbal bullying means using any form of language to cause the other person distress. Examples include using profanities, hurtful language, negatively commenting on a person's appearance, using derogatory terms, or teasing.

Emotional Bullying Emotional bullying involves using ways to cause emotional hurt to another person. This can include saying or writing hurtful things, causing others to gang up on an individual, and purposely ignoring, or spreading rumors.

Social Bullying This includes rolling your eyes or turning away from someone, excluding others from the group, getting others to ignore or exclude, gossiping or spreading rumors, setting others up to look foolish, and damaging reputations and friendships.

Personal Bullying Personal bullying refers to any sort of bullying, done in any manner that is related to a person's gender or sexuality. Including leaving someone out; treating them badly or making them feel uncomfortable because of their sex; making sexist comments or jokes; touching, pinching, or grabbing someone in a sexual way; making crude comments about someone's sexual behavior or orientation; or spreading a sexual rumor.

Racial Bullying This includes treating people badly because of their racial or ethnic background, saying bad things about a cultural background, calling someone racist names, or telling racist jokes.

Religious Bullying This includes treating people badly because of their religious background or beliefs, making negative comments about a religious background or belief, calling someone names, or telling jokes based on his or her religious beliefs to hurt them.

Disability Bullying This includes leaving someone out or treating them badly because of a disability, making someone feel uncomfortable, or making jokes to hurt someone because of a disability. On a physical level, it includes harmful actions such as blocking ramps and elevators, tripping, or tampering with accessible equipment.

The following discussion focuses specifically on six bullying types, and how each type may affect your third and fourth-grade students.

Corresponding lesson plans for classroom exercises and discussion, of each of these six bullying types, are included in the lesson plan section at the end of this chapter.

Cyberbullying   Cyberbullying is any type of bullying that happens online or via any electronic means such as cell phones.

Physical Bullying   Physical bullying always involves physical contact with the other person, this type of bullying can also cause harm to property and belongings.

Verbal Bullying   Verbal bullying means using any form of language to cause the other person distress including teasing.

Emotional Bullying   Emotional bullying involves using ways to cause emotional hurt to another person.

Social Bullying   Social bullying includes behaviors that make others look foolish and can damage reputations and friendships.

Disability Bullying   This includes leaving someone out or treating them badly because of a disability.

The CDC indicates that different types of violence are connected and often share root causes. Bullying is linked to other forms of violence through shared risk and protective factors. Addressing and preventing one form of violence may have an impact on preventing other forms of violence (Figure 1.4).[22]

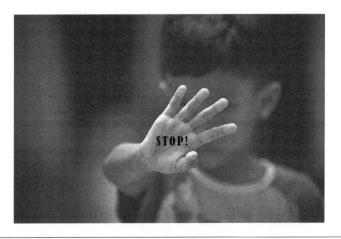

Figure 1.4   Stop bullying.[23]

School bullying is a global concern. As such, UNESCO Member States declared the first Thursday of November, the International Day against Violence and Bullying at School, including Cyberbullying, recognizing that school-related violence in all its forms is an infringement of children and adolescents' rights to education and their health and well-being.[24]

This presents a great opportunity for schools and child-serving organizations to create an activity or learning fair around this day for student, family, and community awareness. UNESCO themes each year and an activity or event may be created around the global theme. By incorporating the theme, students may create poster boards and drawings depicting bullying to display on the school premises such as halls, or in school/community events.

Your Role in Monitoring and Stopping Bullying

If there is an imminent threat of any kind, report the activity to authorities immediately.

Working with your school superintendent and governmental agencies is a starting point for developing a strategy for addressing bullying behavior. Currently, no federal law directly addresses bullying. In some cases, bullying overlaps with discriminatory harassment, which is covered under federal civil rights laws enforced by the U.S. Department of Education (ED) and the U.S. Department of Justice (DOJ). No matter what label is used (e.g., bullying, hazing, and teasing), schools are obligated by these laws to address the conduct when it meets all three criteria below. It is:

- Unwelcome and objectively offensive, such as derogatory language, intimidation, threats, physical contact, or physical violence.
- Creates a hostile environment at school. That is, it is sufficiently serious that it interferes with or limits a student's

ability to participate in or benefit from the services, activities, or opportunities offered by a school; Based on a student's race, color, national origin, sex, disability, or religion.[25]

Every state and most U.S. Territories have some level of Anti-Bullying Laws, Policies, and Regulations. The common components found in state laws, policies, and regulations — which have evolved — include definitions of bullying, defining characteristics that are commonly targeted for bullying behaviors, and detailed requirements for school district policies. A table of state and territory common components is available at Common Components in State Anti-Bullying Laws, Policies and Regulations | StopBullying.gov.[26]

Steps for consideration in developing a school plan for addressing bullying in your school are offered as a place to start. Add what is appropriate in your school and your community, it is an organic process, subject to revision, and enhancement as needed.

1. Ensure you have your school administration's support.
2. Contact your school district to determine what policies, guidance, and support are currently available.
3. Who will be responsible for directing the policy development?
4. Who will be involved in the development process? Teachers and administration only? Include students? Parents and guardians? Law enforcement? Professional service agencies that deal with adolescents and bullying?
5. Establish the objective/s of your anti-bullying policy.
6. Determine the goals you wish to achieve. Consider the SMART goals method — Specific, Measurable, Achievable, Relevant, and Timely
7. Understand and plan your policy's vetting process for fairness and to ensure the policy complies with federal and state laws.
8. What type of mediation will be available to students — both those being bullied and the bully?
9. When will intervention be required? Are there specific triggers for automatically calling law enforcement or a professional agency?

10. What is the process to determine what organizations to contact for referral?
11. Decide how you will implement your anti-bullying policy.
12. What metrics will be used to establish that the policy is working?
13. How frequently will the policy be reviewed for currency and compliance with changing federal and state laws?
14. Who is responsible for managing and maintaining the anti-bullying policy?

These steps will begin the process of ensuring your school is proactive in safeguarding all students from bullying behavior.

As is often the case, many bullying responses include "I was just teasing." or "Don't make a big deal of it, they were just teasing." The line between teasing and bullying may be easily crossed, if not managed properly.

"But I Was Just Teasing." Teasing and Bullying

Sometimes teasing is harmless and playful. Other times it can be used to hurt others. Even playful teasing can hit raw nerves or be misinterpreted, especially when kids struggle with social skills. Many kids tease each other to bond or form relationships. The teasing shows each other they can joke around and still be friends.

But teasing can also be used to communicate the negative. It's often used to establish "top dog" amongst kids — an imbalance of power. Also, what's playful to one child may not feel playful to another. In those cases, teasing can lead to hurt feelings. Another distinct difference between teasing and bullying is that bullying is repeated, it continues with regularity, even after the subject of the bully has indicated it is not appreciated or taken playfully.

With these negatives, why not discourage teasing completely? Like any communication, teasing has its purpose. Some topics that are awkward to raise in serious conversation are easier to raise through teasing.

- Teasing and bullying are different.
- Not all teasing is bad. Sometimes it's playful and helps kids bond.
- When teasing is meant to hurt and is done over and over, it can become bullying.

Teasing can also be fun. Think, for example, of the back-and-forth banter that happens in any romantic comedy.

Verbal bullying is different from teasing. It's not done to make friends or to relate to someone. Just the opposite: The goal is to embarrass the victim and make the bully look better and stronger.

The tricky thing is that bullying may start as teasing. But when teasing is done over and over and is meant to be hurtful or threatening, it becomes bullying.

Unlike kids who are being bullied, kids who are being teased can influence whether it continues or ends. If they get upset, the teaser usually stops.

Sometimes, kids who are trying to tease end up bullying. For example, a child may say something mean-spirited to another, thinking it's playful. This can lead to an argument. Or a child may react angrily to a friendly comment, which may cause other kids to keep their distance (Figure 1.5).

To address these struggles, it's important to teach kids about the rules of conversation. Help kids sort out when teasing is okay and when it becomes hurtful or borders on bullying. One way to do this is by role-playing with them. This lets kids practice a situation where they get teased, don't like it, and need to respond.

Figure 1.5   Teasing girl.[27]

Questions to Ask Kids about Teasing

Maybe you've heard that kids are teasing your child or your student at school. You can ask a few questions to see whether it's good-natured or harmful:

- Are the kids who tease you your friends?
- Do you like it when they tease you?
- Do you tease them back?
- If you told them to stop teasing, would they?
- If you told them that they hurt your feelings, would they say they were sorry?

If the answer to any of these questions is "no" or "I don't know," then it may be a case of negative teasing or even bullying. It's important to find out more.[28]

Children need to learn there is a difference between bullying and teasing, both in person and online. Remember, if there is an imminent threat of any kind, report this to the authorities immediately.

How to respond to bullying situations is a good classroom activity to practice. This practice is a sound way to prepare students for a bullying situation. It may also make the bully aware of their behavior, and that is it neither appropriate nor appreciated. This is a guided practice and is not intended to make anyone feel bad or guilty.

The list indicating the different types of bullying, previously discussed, may be used to create bullying scenarios. Scenarios should not reflect any identifiable student.

Guided responses may include:

- "That is my friend and don't talk to them that way." Shows support and that there are allies.
- "I do not like you saying/doing that. Please stop." It is a firm statement and a request to stop. This is a good approach as the immediate response by the bully and future activity will also assist in determining if it is bullying or teasing.

- "I have asked you to stop before. I am reporting this to the teacher."
- "It makes me embarrassed when you say/do this."

Remember, if there is an imminent threat of any kind, report this to the authorities immediately.

In the following section, we take a closer look at what is good judgment and how we communicate this principle, this concept to young learners. Using good judgment is not only an important life skill but becomes especially important when navigating the oft-times confusing and perilous world of technology. Using good judgment is a primary cyber-safe skill!

Using Good Judgment by Making Good Decisions

Judgment and decision-making is an integral part of discussing character education. As indicated in the introduction to this chapter, "The core mission of character education is to guide students to make thoughtful choices and act with integrity." Creating an environment where students feel safe making choices, empowers the student to feel confident in making future choices.

What Is Good Judgment?

To consider the development of good judgment in children and guiding children in the method to make good decisions is about making good choices. Good judgment starts with knowing how to assess options and the consequences involved in making a choice.

Using good judgment as well as good decision-making is the foundation for ethical and moral decisions the child will continue to make as they get older. The Menninger Clinic, a mental health facility in Houston, states that "graduates get into the most elite colleges but can't handle college life emotionally — and so take a medical leave of absence and come to the Menninger Clinic for treatment and went on to say that all of these students had had insufficient experience making decisions for themselves, handling setbacks, and managing life's temptations independently."

Authors Johnson and Stixrud continue reinforcing the need for early decision-making skills, stating...

> So, while our impulse is to try to keep our kids safe, our priority as they get older should shift to helping them develop the skills — including the decision-making skills — they need to keep themselves safe, along with the confidence that they can trust their judgment and solve problems as they arise. In our view, the time to start supporting kids in making their own decisions isn't when we send them to college — it's when they're young — because we want them to have a ton of practice making and learning from their own decisions...[29]

Making good decisions is about choices. It is easier to make bigger and better decisions when we have the foundation for identifying how to make good choices. Part of helping children make a choice is recognizing that the adult — parent, guardian, or teacher for example — will need to release control of the decision-making process.

Neuroscience News proposes that children as young as six factor in past choices when making moral judgments. Involving children aged four to nine, the study revealed that younger children's judgments were mainly influenced by the actual outcome, whereas older children factored in what could have been done differently.

Through this counterfactual thinking, children from age six began to exhibit more nuanced and mature moral assessments. The findings could help in guiding more effective moral education for young children.

Key Facts:

1. From the age of six, children begin to incorporate past choices into their moral judgments, exhibiting a more mature and nuanced understanding of behavior.
2. In contrast, four and five-year-old children's moral judgments are influenced solely by the actual outcome.
3. This research provides the first direct evidence that children consider counterfactuals in their moral judgments, paving the way for more comprehensive moral education strategies.

A new study published in the journal *Child Development* from researchers at Boston College in Massachusetts (USA), and the University of

Queensland in Australia explores whether four- to nine-year-old children consider past choices when making moral judgments of others. Across two studies, 236 (142 females) children aged four to nine were told stories about two characters who had a choice that led to a good or bad outcome, and two characters who had no choice over a good or bad outcome.

The findings showed that from the age of six, children considered what characters could have done when making judgments of how nice or mean they are behaving and that four and five-year-olds' moral judgments were influenced only by the actual outcome.[30]

Choices — Where to Start

From a toddler choosing what toy to play with to a high school senior selecting a college, children of all ages must make decisions. While some decisions might seem easy to an adult, they could require skills a child may not have developed. And as kids get older, the decisions become more important and more complicated. That's why it's important to help children build decision-making skills from an early age, by giving them lots of practice in a developmentally appropriate way.[31]

Many choices are more important than others in terms of impact and risk, so start with less serious scenarios and alternatives. If the choice made is not inappropriate and is not harmful to the child or to others, start with simple exercises. It is best to start by presenting the student with a limited number of choices rather than open-ended options. As experience and confidence grow by making these less impactful decisions, the student will be able to tackle more complex choices.

Here are a few examples where both the impact and the risk involved in the choice are low. For parents and guardians, a good place to start is to let children choose if they want to go to a friend's party or out to a family lunch. Discussion points include how the friend may feel if you go/do not go to the party, and will the party or the lunch happen again, providing an opportunity to join the activity at a later date? This example and the discussion points may be transferred using school activities.

For educators, give the child choices about solitary activities such as drawing a picture or reading a book. Both are appropriate and neither is harmful to the student or other students. These may be used as good discussion activities on days when inclement weather prevents outside playground activities.

This will take time as it is essentially an individualized activity! If class sizes are large and this level of one-on-one is not available, options may be to select one child each day or week, or alphabetically by name, as determined by frequency available. When rotation through the class roster is completed, begin again. The entire class will learn by observing the process as each child makes their decision. Remember, the student is being presented with two right choices, so there is no wrong answer.

A significant result of decision-making and choices is outcomes and consequences. These can be good or bad, or while using the simplest of scenarios, the choice may have little impact. Selecting either drawing a picture or reading a book is neither good nor bad. As choices are presented, the outcome of each option must be considered and reviewed with the student.

Helping the student make a choice should not be left to a flip of the coin (Figure 1.6).

Figure 1.6 Making good decisions is more than flipping a coin.[32]

Here are some tips from the Wellspring Center for Prevention to help children develop healthy decision-making habits:

Encourage critical thinking.

- Teach children to think critically by asking open-ended questions and encouraging them to weigh the pros and cons of their choices.

Model good decision-making.

- Children learn by example, so be a good role model by making thoughtful decisions and explaining your reasoning to them.

Teach problem-solving skills.

- Help children learn problem-solving skills by encouraging them to identify and evaluate different solutions to a problem.

Foster independence.

- Allow children to make decisions for themselves, within reason, and encourage them to take responsibility for their choices.

Provide information.

- Provide children with accurate and age-appropriate information to help them make informed decisions.

Support resilience.

- Encourage children to bounce back from mistakes and setbacks and teach them that failures are opportunities to learn and grow.[33]

Do Adults Make Decisions Differently Than Children?

Neuroscience News continues summarizing the study indicating that when making moral judgments of past actions, adults often think counterfactually about what could have been done differently.

"Our findings highlight how understanding the choices someone had is an essential feature of making mature and nuanced moral judgments," says Shalini Gautam, a postdoctoral researcher at Boston College.

"It shows that children become able to do this from the age of six. Children younger than six may not yet be incorporating the choices someone had available to them when judging their actions."[34]

Guiding younger children in making decisions provides the tools children need to make balanced and reliable decisions in the future as choices become more difficult and have a greater impact on the child and those around them.

Neuroscience News studies further show that some kids are risk-takers. Others tend to play it safe. Are these differences simply based on personality, or do children's environments help shape their willingness to take a gamble? A new study from researchers in Boston University's Social Development and Learning Lab shows children from different socioeconomic backgrounds make different decisions when placed in the same risky situation.

Peter Blake is a study coauthor and a Boston University College of Arts & Sciences associate professor of psychology. "I hope this study — as well as other future studies by our lab and other people — will change perspectives," Blake says. The research provides evidence that risky decisions in childhood do not always reflect poor judgment or a lack of self-control.

To test human applications of this theory, Blake and his coauthor, Teresa Harvey (GRS'20), built an experiment to see if children's risk preferences would vary by their socioeconomic status and by the size of the offered rewards.

Dozens of children between the ages of 4 and 10 participated in the study, which was conducted at several research sites in Greater Boston, including the Museum of Science. Each child was given the choice to accept a set number of stickers or to spin a wheel for a 50/50 chance of getting even more stickers — or nothing at all.

After some easy practice rounds that ensured participants understood the task, the children were given tougher choices, including a large-reward option (keep four stickers or spin for a chance of getting eight stickers or none) and a small-reward option (keep two stickers or spin for a chance of getting four stickers or none).

While the children participated in the experiment, their parents filled out demographic forms that included questions about the parents' education levels and income.

When the researchers analyzed their data, they found that children from families with lower socioeconomic status were more likely to take a risk and spin the wheel in the large-reward trial than were children from higher-status families. Socioeconomic status made no significant difference in the small-reward trial.

"The kids with lower socioeconomic status, they followed the pattern predicted by the theory," says Blake. "They acted like the hungry fox. They were more likely to take the risk to get the larger reward, and when it came to a lower value reward, they chose the certain option so that they would get something."

The study also showed that boys were more likely than girls to make risky decisions, but gender differences didn't affect the socioeconomic patterns the researchers were interested in. The study showed no age-based differences in risk preference.

Blake says he hopes parents, teachers, and others who see a child making risky choices will pause and consider that such decisions might make sense given the child's circumstances.[35]

Assessing the thought and decision-making method in young children may be a nuanced process. Just as many factors influence adult decision-making, children have many factors influencing them — they just may not be able to articulate or control those factors.

Artificial Intelligence

Introduction

The following discussion provides the reader with a non-technical overview of artificial intelligence (AI)…a technology that has become both a bane and a benefit to teachers since the debut of the most talked about, publicly available AI tool, ChatGPT.

The power and popularity of current and emerging AI tools create an environment that is productive, rewarding, and beneficial to users, while also presenting these same users with unseen risks.

Before we "look under the hood" and talk more about AI, discussing this topic with your students may best begin by simply helping your students understand that machines and technology are a part of our daily lives.

Ask your students, for example, about the devices they use (computers, tablets, etc.), and talk about how technology helps us in various ways.

Defining and explaining technology for third and fourth graders may at times prove challenging; however, it can also be fun, while also being instructive.

One suggested approach to having this discussion with students may proceed as follows…

Technology is like a set of tools and toys that we use to make things easier or more fun.

It includes gadgets like computers, tablets, and robots.

Just like crayons help us draw, technology helps us do cool things and especially learn new stuff!

Figure 1.7 Defining technology for third- and fourth-grade students.[36]

 **Teaching Tip** Hey kids, technology is like a huge collection of tools that help us solve problems and make our lives easier! Imagine Batman's utility belt with all his gadgets, or Dora the Explorer's backpack. Just like them, we have all sorts of cool stuff, like computers, smartphones, and even the microwave that heats your pizza! These tools were invented by very creative people who wanted to help us do things quicker, learn new stuff, and have fun. Technology is all about using science to create helpful inventions, like helping doctors look inside our bodies with special machines or letting us talk to someone far away with just a click. It's pretty awesome, right? (see Figure 1.7).

What Is Artificial Intelligence (AI)?

One of the challenges with defining AI is that if you ask 10 different academics, data scientists, or industry practitioners, you will receive 12 different definitions. Defining AI is a moving target. The folks involved with AI today (which seems like almost everyone) haven't converged yet on exactly what a comprehensive, workable, acceptable definition is.

National Institute of Technology (NIST) Information Technology Lab Director Chuck Romine, referring to the 2019 American AI Initiative, Executive Order 13859 which resulted in the National AI Initiative Act of 2020, provides us with a down-to-earth definition, selected by the authors for its directness, simplicity, and usability when discussing AI with students.

Director Chuck Romine defines an AI system as a system that exhibits reasoning and performs some sort of automated decision-making without the interference of a human.[37]

How Do AI Systems "Learn?"

Just like developing good characteristics in students throughout their educational journey, technology like AI must be taught to demonstrate good characteristics. AI systems that exhibit characteristics like resilience, integrity, reliability, security, robustness, interoperability, and privacy are exhibiting good characteristics. If AI systems are going to be trusted, useful, adopted, and accepted by people without fear, they must exhibit good characteristics and acceptable behavior.

Teaching AI systems to behave acceptably and embrace approved societal good behaviors and norms, like students, must be taught. Teaching AI systems is not too different than teaching students. In the classroom and through coaching, exercises, and lessons, students are presented with examples of desirable or good characteristics and acceptable behavior. These lessons are repeated throughout a student's academic experience.

Training AI Systems

Training AI systems involves a process similar to teaching a computer how to perform specific tasks or make decisions by exposing it to large amounts of data. This process is known as machine learning (ML). At its core, ML is the subset of AI that enables systems to learn and improve from experience without being explicitly programmed. The fundamental concept is to input millions if not billions bytes of data that represent information into the AI system. This input or "feeding" process allows the AI system to recognize patterns, make predictions, or classify information autonomously.

One primary method of training AI is called supervised learning. In this approach, the AI system is provided with a labeled dataset, where each input is paired with the correct output. For instance, if we want the AI to identify pictures of cows and horses, we will present the AI system with a collection of images labeled as either a cow or a horse.

The AI learns by adjusting its internal parameters through repeated exposure to these labeled examples, refining its ability to make accurate predictions. Supervised learning is analogous to a teacher guiding a student with correct answers during a learning process.

Another key training method is unsupervised learning. Here, the AI explores data without explicit guidance, seeking to identify hidden patterns or relationships. Clustering, a common unsupervised learning technique, involves grouping similar data points together. Imagine a scenario where the AI receives a mix of unlabeled pictures and independently discovers that some share characteristics, effectively categorizing them into groups. This unsupervised learning is comparable to a student independently organizing information based on inherent similarities.

Reinforcement learning is a third approach, inspired by behavioral psychology. In this method, the AI, often referred to as an agent, learns to make decisions by interacting with an environment. It receives feedback in the form of rewards or penalties based on the actions it takes. This trial-and-error process enables the AI to learn optimal strategies for maximizing rewards. Reinforcement learning can be compared to a student learning through experimentation and adjusting their behavior based on outcomes.

The training process also involves neural networks, which are computational models inspired by the human brain's structure. Neural networks consist of interconnected nodes (or artificial neurons) organized into layers. During training, the network adjusts the weights assigned to connections between nodes, optimizing its ability to make accurate predictions. The deep learning paradigm involves neural networks with many layers, allowing them to capture intricate features in the data. Visualize this as a student's brain adapting and strengthening connections as they acquire more knowledge and skills over time.

To sum up our review of AI and the various subsets that make up the broader field of AI, we are left with one elementary question… "How does artificial intelligence use data to make informed decisions?"

The answer is uncomplicated and straightforward…AI uses data to make informed decisions through a process called "machine learning." ML is a subset of AI that enables systems to learn from data, identify patterns, take actions, and make recommendations, predictions, or decisions based on that data (see Figure 1.8).

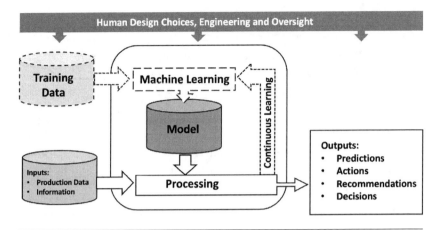

Figure 1.8   AI functional overview.[38]

An Example for Third and Fourth-Grade Students

So, how do we translate this information on how AI is trained to provide us with answers to our questions, into words and concepts understandable to elementary-aged children?

Here is one example of an in-class exercise you may wish to consider and adapt for your students.

Begin your discussion on how we train AI systems by asking your students to imagine that they have a pet parrot, a really smart pet parrot. Then explain that this parrot doesn't just repeat what they say, but it can learn and give them answers to questions they may ask it. That's kind of like what AI is…computer software but, really smart computer software.

To teach the parrot to answer questions, they will have to give it lots of examples. So, explain that they might show their pet parrot a picture of a banana and say "banana." They would show the parrot many pictures of bananas, every day and do this with many pictures and many words, so their pet parrot learns a lot.

AI works the same way. It learns from lots and lots of information — like books, websites, and pictures. People who know a lot about computers help teach AI by giving it rules, kind of like the rules in a game, to help it understand what it learns (see Figure 1.9).

Then, when you ask the AI a question, it uses what it learned, like your smart pet parrot, to give you the best answer it can. It's not

Figure 1.9   Training AI.[39]

perfect, but it gets better the more it learns, just like you get better at a game the more you play.

Providing a clear example will help to solidify the concept in their minds. Begin by explaining that you have a question that you want to ask the AI software. Your question…"What's the biggest animal in the ocean?" Through all of the training that the AI software has been subject to, the AI software remembers all the times it was provided with information about ocean animals. Then, it thinks fast and tells you, "The blue whale is the biggest animal in the ocean!" That is how AI learns.

In summary, training AI systems encompasses various methods, such as supervised learning with labeled datasets, unsupervised learning for discovering patterns independently, and reinforcement learning for decision-making through interaction with an environment. Neural networks, inspired by the human brain, play a crucial role, in adjusting their connections to improve accuracy. The process is analogous to guiding a student through examples, allowing them to learn independently, and refining their decision-making through experience.

What Is Generative AI (GEN-AI)?

The popular GEN-AI tool, ChatGPT 4 was presented with the following request… "Please explain what generative AI is using non-technical terms." ChatGPT 4 responded by providing this definition….

Generative AI is like a virtual artist in your computer. Imagine you tell this artist to paint a picture, write a story, or even compose music. You don't have to give very specific instructions. You can say something like, "Draw me a picture of a beach at sunset," or "Write me a poem about love," and the AI will create it for you.

Asked to explain further, ChatGPT 4 added the following....

This virtual artist isn't just copying things it's seen before; it's using what it has learned from lots of examples to make something new and original. It's a bit like how a chef might learn to cook by trying lots of recipes and then invent a new dish. Generative AI does this with art, words, music, and more, learning from lots of data and then making something unique on its own.[40]

When explaining GEN-AI to your students, simple is best .... GEN-AI is like a smart computer artist who can create new pictures or stories by learning from old ones.

Caution: AI Ahead

One risk associated with GEN-AI is exactly what makes GEN-AI "intelligent," the use of excessively large amounts of data. Data that can come from both known (substantiated) and unknown (unsubstantiated) sources.

Due to the vast amounts of data that GEN-AI models have access to, the ability for the generation either intentional or unintentional, of harmful content is highly probable. Content that may inaccurately attribute comments to someone who never made the statement or alter images to deceive the viewer into believing that the image represents an authentic scene, person, or event.

In the field of AI and ML, this is referred to as hallucination. The production of plausible seeming but factually incorrect output by a generative AI model that purports to be asserting the real world, however, the AI system provides an answer that is factually incorrect, irrelevant, or nonsensical, because of limitations in its training data and architecture.

The potential for GEN-AI models to increase efficiency and productivity, make better decisions, improve the "speed" of business, reduce human error, and improve services comes with ethical concerns and risk exposures inherent in AI technologies and GEN-AI models.... accountability and transparency, the potential to create and to distribute harmful content, privacy exposure, loss of data protection, trust and explainability concerns, job displacement, unemployment, and economic disruption, and data biases.

Throughout their academic journey, students learn about being an ethical person, and that responsibility and trust represent good characteristics and acceptable behavior. These lessons need to be reinforced and discussed with students in terms related to the world of technology. This is even more important now with the advent and presence of AI as a tool, a resource gaining increasing popularity in all sectors of society, including academia, and the increasing availability for student access and use.

Just as students learn what ethical and responsible behavior is, that trust is earned, how to trust, and that not everyone can be trusted, students must be provided with instruction and guidance on using AI ethically, and responsibly and that AI cannot always be trusted (meaning that AI, as it exists currently, will not always provide the correct answer).

AI Hallucinations

The fact that AI cannot and does not provide a correct response 100% of the time (yet) has resulted in the introduction of a new term to our social lexicon...Hallucinations, specifically AL Hallucinations.

In the constantly evolving field of AI, hallucinations are incorrect or misleading results that AI models generate. These errors can be caused by a variety of factors, including insufficient training data, incorrect assumptions made by the model, or biases in the data used to train the model. AI hallucinations can be a problem for AI systems that are used to make important decisions, such as medical diagnoses or financial trading.

AI models are trained on data, and they learn to make predictions by finding patterns in the data. However, if the training data is incomplete or biased, the AI model may learn incorrect patterns. This can lead to the AI model making incorrect predictions, or hallucinating.

AI hallucinations can take many different forms. Some common examples include:

- False Positives: When working with an AI model, it may identify something as being a threat when it is not. For example, an AI model that is used to detect fraud may flag a transaction as fraudulent when it is not.

- False Negatives: An AI model may fail to identify something as being a threat when it is. For example, an AI model that is used to detect cancer may fail to identify a cancerous tumor.[41]

Unethical Uses of AI

If the past several months are any indicator of the widespread acceptance of GEN-AI as a tool for the masses, students as well as adults must be taught to engage with and to use AI ethically.

The ramifications of the unethical use of AI are numerous and as AI matures the list of potential consequences of unethical AI grows as well (see Figure 1.10).

Currently, the most significant issues of unethical AI are:

- Privacy Erosion: The unethical use of AI poses a significant threat to privacy. As AI systems gather and analyze vast amounts of personal data, there is a risk of unauthorized access, data breaches, and misuse.
- Bias and Discrimination: Unethical AI practices can perpetuate and even exacerbate societal biases. If the training data used to develop AI algorithms contain biases, the AI systems may produce discriminatory outcomes.
- Security Threats: The unethical use of AI can introduce new security threats, as malicious actors leverage AI to enhance the sophistication of cyber-attacks.
- Job Displacement and Economic Inequality: The deployment of AI in the workforce, without ethical considerations, can lead to job displacement and economic inequality.
- Autonomous Weapons and Warfare: The unethical use of AI in the development of autonomous weapons raises profound ethical and humanitarian concerns.
- Deepfakes: A type of AI used to create convincing images, audio, and video hoaxes. The term describes both the technology and the resulting bogus content and is a hybrid of deep learning and fake information. The greatest danger posed by deepfakes is their ability to spread false information that appears to come from trusted sources.[42]

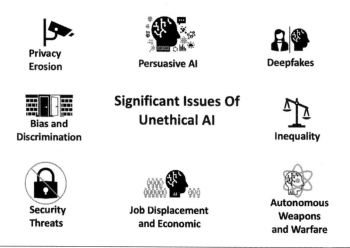

Figure 1.10 Significant issues of unethical AI.[43]

- Persuasive AI: Refers to technology, like robots or computers, that is designed to influence or convince people to think or act in certain ways through words, images, or actions.
- Inequality: In the context of unethical uses of AI refers to the unjust and disproportionate distribution of benefits, opportunities, or resources among individuals or groups due to biased algorithms, discriminatory practices, or unequal access to AI-driven technologies.

Addressing the most significant issues of unethical AI requires a global recognition of the risks that can be inherent within AI and an effort to establish ethical frameworks, regulations, education programs, and responsible AI practices.

Teach Your Children Well

The message contained within the lyrics of the Crosby, Stills, Nash & Young song *Teach Your Children* is as relevant today as when Graham Nash wrote it in 1970...we all have a responsibility to pass on our values, beliefs, and dreams to the next generation so that they can build a better world.[44]

AI will have a significant impact on the world in which our students live, learn, and grow. Teaching students to use AI ethically and responsibly is preparing them to manage AI as adults and use AI to build a safer, better, world.

Teaching students about ethical behavior and behaving responsibly can at times be challenging. For educators, applying these life skills to AI adds to this challenge.

Educators may wish to consider the following when discussing the ethical and responsible use of AI with their students.

1. Emphasize that AI is designed to help people and make their lives easier. Provide examples such as smart assistants like Siri or Alexa, which can answer questions or play music. Explain that AI is a tool that can assist us in various ways.

2. Teach the importance of privacy by explaining that AI works with information and data. Emphasize that it's essential not to share personal information online or with AI devices without asking a grown-up for permission.

3. Discuss the idea that AI should be fair and treat everyone the same. Use examples like choosing games or activities, highlighting that AI should not favor one person over another.

4. Stress the importance of being truthful when interacting with AI. Explain that providing accurate information helps AI learn and do its job better. Encourage students to use AI responsibly and not to trick it.

5. Explain that AI learns from people, just like students learn from their teachers. Encourage a collaborative mindset by discussing how we can teach AI good things and help it become smarter in positive ways.

6. Introduce the concept of being a good digital citizen. Teach the importance of using AI and technology in a kind and respectful manner. Discuss how to communicate online and treat others with kindness, even when using AI.

7. Emphasize that it's okay to ask for help when using AI. Encourage students to talk to their parents, their teacher, or other grown-ups if they have questions about AI or if something online makes them feel uncomfortable.

8. Clarify that AI has limits and may not always have the right answers. Teach students to have realistic expectations and that it's okay if AI doesn't know everything.

9. Discuss the importance of balancing screen time and offline activities. Teach students that while AI can be fun and helpful, it's essential to have a balance by playing outside, reading books, and spending time with family and friends.

By incorporating these concepts into age-appropriate discussions and activities, the educators of our third- and fourth-grade students can lay the foundation for ethical and responsible AI use, promoting positive digital behavior among young learners.

Chapter 2, "Nurturing Digital Citizens: Cyber Safety for Early Learners," provides the educator with insights into discussing and teaching students the necessity and importance of cyber safety as grade schoolers and as they grow in an ever-changing, technology-infused digital world.

Discussing Artificial Intelligence with Third- and Fourth-Grade Students

AI Chatbots

From ChatGPT to autonomous vehicles, AI is one of the most exciting (and controversial) technology trends happening in the 21st century. The growth of AI-based tools for use in both education and recreation provides students with an ample selection of choices...some good, some not-so-good. Teaching grade schoolers the difference and how to use these tools safely and properly early in their educational journey will prepare them to be cyber-safe and lifelong vigilant digital citizens.

Explaining the concept of cyber safety, AI, and a chatbot's relationship to both, in the same lesson can be a challenge.

After discussing the general concepts of AI and working through one or more of the suggested lesson plans included at the end of this chapter, move on then to explain exactly what a chatbot is, in terms understandable to third and fourth-grade students.

Due to the vast amount of publicity that ChatGPT has (and continues to) receive, it may be best to begin any discussion about AI and bots with a brief explanation of exactly what is GPT.

GPT, or Generative Pre-trained Transformer, is an AI model developed by OpenAI. It's designed to generate human-like text based on the input it receives. GPT achieves this by pre-training on vast amounts of text data, learning patterns, and structures of language.

This enables it to generate coherent and contextually relevant text in response to prompts or questions. GPT has been used for various applications, including text generation, language translation, and content summarization.

For the educator, at the most basic level, a chatbot is a computer program that simulates and processes human conversation (either written or spoken), allowing humans to interact with digital devices as if they were communicating with a real person. Chatbots can be as simple as rudimentary programs that answer a simple query with a single-line response, or as sophisticated as digital assistants that learn and evolve to deliver increasing levels of personalization as they gather and process information.

Driven by AI, automated rules, natural language processing (NLP), and ML, chatbots process data to deliver responses to requests of all kinds.

There are two main types of chatbots.

- *Task-oriented* (declarative) chatbots are single-purpose programs that focus on performing one function. Using rules, NLP, and very little ML, they generate automated but conversational responses to user inquiries. Interactions with these chatbots are highly specific and structured and are most applicable to support and service functions — think robust, interactive FAQs
- *Data-driven and predictive* (conversational) chatbots are often referred to as virtual assistants or digital assistants, and they are much more sophisticated, interactive, and personalized than task-oriented chatbots. These chatbots are contextually aware and leverage natural-language understanding (NLU), NLP, and ML to learn as they go. They apply predictive intelligence and analytics to enable personalization based on user profiles and past user behavior.[45]

Chatbots have a wide range of applications and can be used in various industries. Some common uses of chatbots include:

- Customer Service: Chatbots can handle customer inquiries, provide support, and assist with common issues, reducing the need for human intervention.
- Information Retrieval: Chatbots can provide information on a wide range of topics, such as weather updates, news, and general knowledge.

- Personal Assistants: Chatbots can act as virtual assistants, helping users with tasks like scheduling appointments, setting reminders, and making reservations.
- E-commerce: Chatbots can assist users in finding products, making recommendations, and completing purchases.
- Education: Chatbots can be used in educational settings to provide personalized learning experiences, answer questions, and assist with homework.

Discussing the concept of chatbots with third and fourth-grade students requires a different, less technical, straightforward approach. An example of this conversation would follow something like this....

Imagine a chatbot as a really smart robot that learns how to talk by reading lots and lots of books. It learns what words go together and how to make sentences that make sense. Then, when you ask it a question or give it a topic, it can use what it learned to give you an answer or tell you something interesting! It's like having a super clever friend who knows everything about lots of different things!

Chatbots can help you with all sorts of things, like answering questions, playing games, or even telling jokes! So, it's like having a clever friend inside your device who's always ready to chat with you! (see Figure 1.11)

Figure 1.11   What is a chatbot?[46]

Using chatbots is now forever ingrained in the academic learning process and online recreation pursuits. Given this reality, the risks associated with chatbots must be discussed with your students. Mentor them to always be vigilant and to stay cyber-safe.

The following risks are the most prominent associated with chatbots and should be reviewed with students:

- Privacy: Sometimes chatbots might ask for personal information like the student's name, age, or where the student lives. It's important to remind students not to share too much personal information with chatbots no matter how friendly the chatbot is, just like the student wouldn't share personal information with a stranger.
- Stranger Danger: Remind students that just like in real life, not everyone they will talk to online is who they say they are. Some people might pretend to be someone else to trick them. So, it's essential to be cautious and not share personal information or meet up with anyone they've only talked to online.
- Inappropriate Content: Chatbots can sometimes share links or information that's not suitable or makes them feel uncomfortable. Remind students that if they see or read something that makes them feel uncomfortable or worried, it's essential that they tell a trusted adult immediately.
- Cyberbullying: Sometimes, people might use chatbots to say mean or hurtful things to others. If someone is being unkind to them or someone else online, it's crucial to talk to a grown-up that they trust about it.
- Online Safety Rules: It is a good reminder to stress that students should always follow the rules that their parents, trusted adult, or their teachers have set for using the Internet. Things like not talking to strangers online, being kind to others, and telling an adult if something seems wrong are all essential cyber-safety rules to remember.

Student-Friendly Chatbots

Some of the more well-known and useful GPT tools that you may wish to review with your students include but, are certainly not limited to the following...

- ChatGPT Kids: An AI chatbot designed specifically for children, providing age-appropriate responses and engaging conversation.
- Storyteller GPT: A tool that helps children generate creative stories by providing prompts and suggestions.
- AI Dungeon: A text-based adventure game powered by GPT, where kids can create and explore their own interactive stories.
- Talk to Transformer: An online tool where kids can input prompts or questions to generate text-based responses from GPT.
- Copy.ai: A platform that generates text for various purposes, including writing stories, creating poems, or generating ideas for school projects.

These are examples of GPT tools that can be fun and educational for children, allowing them to interact with AI in a cyber-safe and age-appropriate way. Any of the GPT tools identified above could certainly be examined as an in-class learning exercise or combined with one of the chapter's lesson plans, an assignment to take home and complete with a trusted adult.

Generative AI Alternatives

Numerous generative AI tools are available for educators to explore, offering a wealth of resources to enhance classroom activities, lessons, and assignments. These tools open up avenues for creative exploration, allowing students to delve deeper into AI capabilities and the potential of these technologies. Additionally, educators may find certain tools beneficial for their own personal and professional growth.

The following list of alternative generative AI tools is incomplete and will probably never be up-to-date. The amount of AI software released daily is staggering. In 2023, more than 10,000 AI tools were released for public use.[47] Given the number of AI start-ups, it is hard to distinguish exactly how many AI tools exist at any given moment, but the total number is probably in the hundreds of thousands.[48]

Listed here are some generative AI tools that you may wish to discuss and explore with your students. Creating either in-class group activities or take-home assignments using one or more of these tools will reinforce learning, provide opportunities to practice proactive cyber-safe skills, and broaden the student's awareness of the many AI tools available to assist them in their academic studies.

Generative AI Tools

Writing Assistants

- AI-Writer: an AI writing assistant that helps reword and remediate an existing piece of content, creates unique article drafts, provides citation lists, and summarizes SEO competitors (a competitor, in the context of SEO, is a website or webpage that is competing with your target page for ranking in the search engine results), and creates SEO-optimized content from Google. SEO (search engine optimization) is the process of improving the quality and quantity of website traffic to a website or a web page from search engines
- Bard: Google's conversational AI tool. Bard aims to improve the way people search for and retrieve information.
- ChatSonic: is a dialogue-focused content generator from Writesonic built on top of GPT-4 with added features. Users can choose from 16 different personas to chat with, including poets and accountants.
- Claude: is a conversational AI tool by Anthropic, which is integrated into Slack to summarize threads and answer questions.
- Docs: Google's cloud-based, collaborative word processor with AI features to generate, summarize, and brainstorm text.
- Freshchat: is a chatbot platform that can be used in education. It can engage in dialogue with students, identify gaps in learning, provide relevant information, answer questions, and suggest alternative strategies. Freshchat can act as a virtual teaching assistant, reducing the teacher's workload and helping students get back on track.
- Grammarly: is the mainstream spell- and structure-checking app. It assists users in keeping their grammar on point, lets writers adjust their tone, and suggests shortcuts to simplify wordy or complex phrases.
- Jasper AI: a long-form AI copywriting tool and article generator that includes more than 50 content generation templates in 25 global languages.
- Koala: is an AI chatbot that can assist students and teachers with generating long-form written pieces such as essays and blog posts. It offers references and sources for its data and is

built on advanced AI technology. Koala can understand variations in user input and pick up on conversational contexts, making it a helpful tool for writing assignments.

- Magic Write: is an AI text generator for Canva Docs. Users can prompt Magic Write to brainstorm, generate outlines, and generate content ideas.
- Perplexity AI is an AI-powered search engine. A unique feature of Perplexity is that it can provide sources to back up the answers that it has generated. It functions more as a search engine than an original writer but still generates original content.
- Rytr: an AI content generator that lets content creators specify a content use case, tone of voice, and keywords.
- Spellbook: is writing software by Rally that is designed to help lawyers with legal drafting. Spellbook can draft new contract clauses, list common negotiation points based on the contract, and create contract summaries.
- Tome: utilizes advanced language models to generate text based on user input or prompts. Tome can be used for various purposes such as brainstorming ideas, generating story outlines, or even crafting entire passages of text.
- Wordtune: assists users in improving their writing by providing suggestions and alternatives for phrases and sentences by offering suggestions for rewriting sentences to enhance clarity, coherence, and style, helping users refine their writing quickly and effectively.

Image Design and Development

- Craiyon: produces a batch of AI-generated images in response to a text prompt.
- DALL-E: is OpenAI's image generator that creates images and art from a simple text prompt.
- Image Creator: is designed to produce images based on user inputs or predefined parameters in the creation of graphic design, art creation, or even generating visual content for websites and presentations.
- ImageFX: is Google's innovative AI-powered image generation tool, enabling users to create images from simple text prompts.
- Stable Diffusion: is stability AI's generator that can create photorealistic images from an inputted text.

Music

- Amper Music: is a generator that creates music from prerecorded samples. The software can be used to match music to video.
- Dadabots: is a generative neural network that creates a constant live stream of artificial music in different genres and raw audio neural networks that imitate bands.
- Soundraw: is an AI music generator that can be used to generate royalty-free background music.

Others

- Bing: while not a generative AI tool; it is a search engine developed by Microsoft. Bing provides users with the ability to search for information, images, videos, and news on the Internet. It utilizes algorithms to index web pages and deliver relevant search results based on user queries.
- Character AI: focuses on creating lifelike characters for various digital applications by generating dialogues, personalities, and backstories for virtual characters in games, simulations, or interactive storytelling platforms. Character AI provides developers and storytellers with a powerful tool for crafting compelling narrative experiences in digital environments.
- Creo: is a computer-aided design system that uses generative design, enabling expedited design of physical objects. It also optimizes designs based on material and manufacturing requirements.
- Grok: is xAI's — Elon Musk's AI startup — version of ChatGPT. It has Internet browsing capability and can answer users based on up-to-date information on the web.
- Jasper Chat: focuses on creating conversational experiences designed to engage users in natural and fluid conversations, simulating human-like interactions. It can be utilized in various applications such as interactive storytelling platforms.
- OpenAI GPT-3 Playground: allows users to interact with and explore the capabilities of the GPT-3 language model. Users can input text prompts and receive generated responses from GPT-3 in real time, enabling experimentation with various

language tasks such as writing, summarization, translation, and more. The playground provides a user-friendly interface for developers, researchers, and enthusiasts to understand, test, and leverage the power of GPT-3 for a wide range of applications.

- Pi is a chatbot designed to serve as the user's personal assistant.[49]
- PepperType is specifically tailored for text generation tasks including content creation, creative writing, and text summarization.
- Perplexity.ai: has access to the Internet and current events and provides prompt suggestions to get chats started.
- Replika: is designed to simulate conversational interactions with users effectively mimicking human conversation. It serves as a virtual companion, aiming to foster meaningful connections and support mental well-being through its conversational interface.
- Socratic: children can type in any question they may have about what they are learning in school and Socratic will generate a conversational, human-like response with fun unique graphics to help break down the concept.
- YouChat: outputs an answer to anything you input including math, coding, translating, and writing prompts. This chatbot cites sources from Google, which ChatGPT does not because ChatGPT doesn't have Internet access.

Summary

This chapter emphasizes the significance of character education for third and fourth graders, focusing on developing ethical qualities that support academic achievements. It highlights the role of character education in fostering a positive learning environment, improving social skills, and encouraging ethical decision-making.

The chapter outlines various components of character education, including values and virtues, ethical decision-making, SEL, and community involvement. It discusses the benefits of character education in creating responsible, respectful, and compassionate individuals who contribute positively to society.

The chapter concludes with the importance of integrating character education into educational programs to support full student development.

LESSON PLANS

Grade 3

TOC Title: G3 Acceptable Behavior
  Lesson Title: Acceptable Behavior
  Grade Level: 3
  Duration: 50–55 minutes (Split into two lessons if needed. The activities may be split into two sessions. When presented in two sessions a 5-minute refresher is suggested for session 2 rather than a full introduction to the topic, and each activity may be presented separately in one session.)

Objective:

- The students will know the importance of acceptable behavior.
- The students will be able to exhibit appropriate conduct in various settings.

Suggested Materials:

- Whiteboard and markers or a projector for visuals.
- Markers.
- Poster Board.
- Colored pencils.
- Construction paper.
- Glue sticks.
- Scissors.
- Books on acceptable behavior (suggestion: Enemy Pie by Derek Munson or similar age-appropriate title covering acceptable behavior).
- School-appropriate magazines or printed images.
- Worksheet for assessment (sample included below).

Procedure

Introduction (10 minutes):

a. Whole group: introduce the topic of acceptable behavior.
b. Lead a discussion about what it means to behave appropriately in different settings such as school, home, and public places.
c. Write important key ideas on whiteboard or chart paper.

Read Aloud (15 minutes):

    a. Read a picture book related to acceptable behavior (suggestion above).

    b. Discuss the characters' behaviors and ask students to identify how those behaviors relate to acceptable and unacceptable behavior.

Activity (15 to 20 minutes) Creating a Behavior Poster:

    a. Divide the class into small groups.

    b. Provide each group with a poster board, markers, and magazines or printed images.

    c. Instruct them to create a poster illustrating examples of acceptable behavior in different settings.

    d. Encourage creativity and collaboration within groups.

    e. After completion, allow each group to present their poster to the class and explain their illustrations.

Conclusion (5 minutes):

    a. Summarize the key points about acceptable behavior discussed throughout the lesson.

    b. Encourage students to apply what they've learned in their a. daily lives.

    c. Emphasize the importance of respect, kindness, and consideration toward others.

Assessment: (5–10 minutes):

    a. Give the assessment worksheet to each student.

    b. Allow students to complete the worksheet independently.

    c. Review the answers together as a class to reinforce learning and clarify any misconceptions.

Homework (Optional):

    a. Encourage students to share what they learned about acceptable behavior with their families.

Sample Assessment Worksheet:

Acceptable Behavior Assessment Worksheet

Name: _____ Date: _____

Instructions: Read each of the following carefully. Decide if the behavior described is acceptable or unacceptable. Circle the correct answer.

1. During recess, someone wants to play on the swings. They see that all the swings are occupied and wait for their turn.
   - Acceptable
   - Unacceptable
2. In the classroom, a student accidentally drops his pencil. He quickly apologizes to the teacher and picks it up without disturbing others.
   - Acceptable
   - Unacceptable
3. At lunchtime, someone finishes their meal and throws away the trash, and his empty juice box misses the trash can. He sees what happened but doesn't pick it up off the ground.
   - Acceptable
   - Unacceptable
4. During a group project, a girl disagrees with her classmates' ideas. Instead of arguing, she listens to their viewpoints and suggests a compromise.
   - Acceptable
   - Unacceptable
5. A student is waiting in line to use the computer in the school library. Before it is his turn, he pushes past another student to get to the computer first.
   - Acceptable
   - Unacceptable

TOC Title: G3 Cyberbullying
  Lesson Title: Cyberbullying
  Grade Level: 3
  Duration: 50 minutes (Split into two lessons if needed. The activities may be split into two sessions. The lesson plan presented in two sessions will require only a 5–10-minute refresher in the second session rather than a full introduction to the topic and the safety rules. The poster activity may be presented separately in the second session.)

Objective:

- Students will understand what cyberbullying is, recognize its various forms, and learn strategies to prevent and respond to cyberbullying.

Suggested Materials:

- Projector and screen.
- Internet safety videos for kids (suggested: Safer Kids Online Hey Pug! — Cyberbullying https://www.youtube.com/watch?v=RnPJVqMy_00) or another video that addresses the topic at an age-appropriate level.
- Chart paper and markers.

Procedure

Introduction (5 minutes):

a. Ask students if they know what bullying is. Discuss different types of bullying they may have heard about or experienced.
b. Introduce the concept of cyberbullying. Explain that it is a form of bullying that occurs online or through electronic devices.

Discussion on Online Safety Rules (10 minutes):

a. Define cyberbullying as the use of digital devices to intentionally hurt, harass, or embarrass someone.
b. Discuss different forms of cyberbullying.

c. Leaving mean comments or messages.

d. Spreading rumors online.

e. Posting embarrassing photos or videos.

f. Intentionally excluding someone from online groups or games.

Video and Discussion (10 minutes):

a. Show an age-appropriate educational video (option suggested above) that reinforces cyber safety concepts.

b. Discuss important topics from the video.

Activity: Cyber Safety Poster (20 minutes):

a. Divide class into groups (2–4 students per group).

b. Provide art supplies and ask students to create posters that promote a positive online environment and discourage cyberbullying.

c. Encourage students to include slogans, positive messages, and illustrations.

d. Display posters around the classroom.

Conclusion (5 minutes):

a. Revisit key points about cyberbullying, its various forms, and the prevention strategies.

b. Ask students to reflect on what they've learned and share one thing they can do to prevent cyberbullying.

Assessment:

a. Participation in group discussions and activities.

b. Understanding demonstrated through responses to scenarios.

c. Creativity and effort put into anti-cyberbullying posters.

Homework (optional):

a. Have students talk to their parents or guardians about what they learned and discuss family rules for online behavior.

TOC Title: G3 Physical Bullying
  Lesson Title: Physical Bullying
  Grade Level: 3
  Duration: 50 minutes (Split into two lessons if needed. The activities may be split into two sessions. The activity presented in two sessions will require only a 5-minute refresher in the second session rather than a full introduction to the topic, and the story and the art activity may be presented separately in each session.)

Objective:

- The students will know what physical bullying is as well as strategies to deal with it.

Suggested Materials:

- Whiteboard and markers.
- Markers, crayons, or colored pencils.
- Small slips of paper.
- Storybook about bullying. Suggestion: Say Something! by Peter H. Reynolds or another book that addresses the topic at an age-appropriate level.

Procedure

Introduction (10 minutes):

a. Start with a class discussion on feelings and emotions. Ask students to share times when they felt happy, sad, scared, or angry.
b. Bring up the term "bullying" and ask students for a definition. Write student responses on the whiteboard.
c. Explain that there are different types of bullying, and today the focus is on physical bullying. Ask for examples and write them on the board.

Story (10 minutes):

a. Read a story on the topic of bullying, such as Say Something! by Peter H. Reynold or another age-appropriate book on the topic. Discuss the story and how the ideas in it can be used to deal with bullying. Emphasize the importance of kindness and empathy.

Activity (20 minutes) Bully-Free Zone:

a. Explain that the classroom is going to be a "Bully-Free Zone." Discuss the rules for treating each other with kindness and respect.
b. Have each student decorate a small slip of paper with a drawing or message about kindness. Collect these slips to create a "Kindness Wall" in the classroom.

Conclusion (10 minutes):

a. Recap the main points of the lesson: what bullying is, specifically physical bullying, and the importance of treating each other with kindness.
b. Remind students about the "Bully-Free Zone" in the classroom and encourage them to practice kindness every day.

Assessment:

a. An Informal assessment of student participation during class discussion and examples presented, and a review of the kindness messages created by the students.

TOC Title: G3 Verbal Bullying
   Lesson Title: Verbal Bullying
   Grade Level: 3
   Duration: 55 minutes (Split into two lessons if needed. The activities may be split into two sessions. The activity presented in two sessions will require only a 5-minute refresher in session one rather than a full introduction to the topic, and each activity may be presented separately in each session.)

Objective:

- The students will know and be able to identify verbal bullying.
- The students will understand the impact of verbal bullying.
- The students will know and be able to employ strategies to respond to verbal bullying.

Suggested Materials:

- Whiteboard and markers.
- Chart paper.
- Index cards.
- Colored pencils or markers.
- Storybook depicting verbal bullying. Suggestion: <u>Spaghetti in a Hot Dog Bun</u> by Maria Dismondy or another book that addresses the topic at an age-appropriate level.

Procedure

Introduction (10 minutes):

a. Open with a class discussion on feelings. Ask volunteers to share a time when they felt happy, sad, or angry.
b. Bring up the term bullying. Ask students if they have experienced Bullying. Write responses on the whiteboard.

Activity 1 (15 minutes):

a. Read a story that shows verbal bullying: book suggested above.
b. Discuss how the character(s) felt and how words can hurt.
c. Lead the class in a discussion on the emotional impact of verbal bullying. Talk about empathy and how it can help.

Activity 2 (15 minutes):

a. Brainstorm with the class strategies to respond to verbal bullying. Write ideas on chart paper or whiteboard.
b. Hand out index cards and ask students to write or draw one strategy they can use when experiencing or witnessing verbal bullying.

Conclusion (15 minutes):

a. Ask students to share their strategies.
b. Discuss why it is important to support each other and create a positive, inclusive environment.
c. Encourage students to report any instances of bullying to a teacher or trusted adult.

Assessment:

a. An informal assessment of student learning can be made based on the students' strategies and discussion.

TOC Title: G3 Emotional Bullying
  Lesson Title: Emotional Bullying
  Grade Level: 3
  Duration: 50 minutes (Split into two lessons if needed. The activities may be split into two sessions. The activity presented in two sessions will require only a 5-minute refresher in the second session rather than a full introduction to the topic, and the read aloud and discussion, and the art activity may be presented separately in each one session.)

Objective:

- The students will be able to define and recognize emotional bullying.
- The students will understand the impact of emotional bullying.
- The students will be able to employ strategies that address emotional bullying.

Suggested Materials:

- Whiteboard and markers.
- Picture book that deals with emotional bullying. Suggestion: My Secret Bully by Trudy Ludwig or another book that addresses the topic at an age-appropriate level.
- Index cards.
- Colored pencils, crayons, or markers

Procedure

Introduction (15 minutes):

a. Whole group: ask students to share different emotions they have experienced and how those emotions make them feel. Write their responses on the whiteboard.
b. Define emotional bullying: Write the term "emotional bullying" on the board and ask students for their thoughts on what it might mean. Discuss and come up with a simple definition together, such as "hurting someone's feelings on purpose, again and again."

Read Aloud (10 minutes):

a. Read a story such as *My Secret Bully* or another picture book that deals with emotional bullying.

b. Lead a class discussion on what emotional bullying looks like. Ask students for examples and write them on the whiteboard.

Activity (15 minutes) Empath Exercise:

a. Give each student an index card and ask students to write down an emotion they have felt recently. Collect the cards, shuffle them, and redistribute them. Students read the emotion on the card they receive and discuss with a partner or small group about how they could support someone feeling that way.

Conclusion (10 minutes):

a. Ask students to share something they have learned about emotional bullying.

b. Discuss how understanding and addressing emotions helps to create a positive and supportive classroom environment.

Assessment:

a. An Informal assessment of student participation through observation during class activities and discussions.

TOC Title: G3 Social Bullying
  Lesson Title: Social Bullying
  Grade Level: 3
  Duration: 50 minutes (Split into two lessons if needed. The activities may be split into two sessions. The activity presented in two sessions will require only a 5-minute refresher in the first session rather than a full introduction to the topic, and each activity may be presented separately in each session.)

Objective:

- Students will be able to define social bullying.
- Students will understand the impact of social bullying on individuals.
- Students will explore strategies to prevent and respond to social bullying.

Suggested Materials:

- Video <u>Social Bullying</u> by Dunk-a Bully https://www.youtube.com/watch?v=9Gotfj08YZ8 or another video that addresses the topic at an age-appropriate level.
- Art supplies (markers, colored pencils, paper).

Procedure

Introduction (10 minutes):

a. Begin with a class discussion about the importance of treating each other with kindness and respect.
b. Define social bullying: explain that it involves actions intended to hurt someone's social status or relationships on purpose.
c. Show a short video that addresses social bullying ("Social Bullying" suggested above).

Activity 1 (15 minutes):

a. Students help create a list on the whiteboard of good friendship behaviors.

b. Discuss social bullying behaviors and why they are hurtful. Behaviors may include excluding others, spreading rumors, making someone feel left out, etc.

c. Brainstorm examples of positive and negative behaviors related to friendships.

Activity 2 (15 minutes):

a. Distribute art supplies to students.

b. Students draw a picture illustrating empathy and kindness.

c. Students share their drawings with the class. Discuss how empathy can prevent social bullying.

Conclusion (10 minutes):

a. Review key points discussed during the lesson emphasizing the negative impact of social bullying on individuals.

b. Discuss strategies to prevent social bullying such as being inclusive, standing up for others, and reporting incidents to teachers or trusted adults.

Assessment:

a. Informal assessment of student understanding can be done during the activity as students draw, write, and talk about their empathy and kindness drawing.

Homework (Optional):

a. Encourage students to share what they learned in the video about empathy and kindness with their friends and family.

TOC Title: G3 Disability Bullying
   Lesson Title: Disability Bullying
   Grade Level: 3
   Duration: 55 minutes (Split into two lessons if needed. The activities may be split into two sessions. The activity presented in two sessions will require only a 5-minute refresher in the second session rather than a full introduction to the topic, and the read aloud and discussion, and the art activity may be presented separately in each one session.)

Objective:

- Students will be able to define disability bullying and understand its impact on people.
- Students will recognize the importance of empathy and inclusion.
- Students will understand strategies to prevent and respond to disability bullying.

Suggested Materials:

- Whiteboard and markers or a projector for visuals.
- Picture books on disability awareness and inclusion such as We're All Wonders by R.J. Palacio or another book that addresses the topic at an age-appropriate level.
- Drawing materials (paper, crayons, and markers).

Procedure

Introduction (10 minutes):

a. Begin with a discussion on what bullying is and ask students to share their thoughts. Write their responses on the whiteboard.
b. Introduce the concept of disability bullying, explaining that it involves mistreating or excluding someone based on their differences related to a disability.
c. Discuss the emotional impact of disability bullying on individuals and highlight the importance of treating everyone with respect and kindness.

Read Aloud and Discussion (15 minutes):

a. Read a story that addresses disability bullying and inclusion (option suggested above).

b. Discuss the story with the students. Emphasizing the importance of understanding and embracing differences.

c. Encourage students to share their thoughts and feelings about the characters' experiences.

Art Activity (20 minutes):

a. Distribute art drawing supplies.

b. Have students create drawings that depict inclusion in the classroom.

c. Encourage them to include different abilities and talents in their drawings.

d. Display the drawings in the classroom to promote a positive and inclusive atmosphere.

Conclusion (10 minutes):

a. Review key points of the lesson.

b. Emphasize the importance of empathy, understanding, and inclusion in creating a supportive classroom environment.

c. Encourage students to share what they learned about cyber safety with their families.

Assessment:

a. Evaluate students' participation in discussions, and how their drawing activity connects with the message of the lesson.

TOC Title: G3 Good Judgment
  Lesson Title: Good Judgment
  Grade Level: 3
  Duration: 50 minutes

Objective:

- The students will understand what good judgment is.
- The students will demonstrate using good judgment in everyday situations.

Suggested Materials:

- Whiteboard and markers.
- Picture books on decision-making and judgment (Suggestions: <u>What Should Danny Do?</u> by Adir Levy and Ganit Levy, <u>The Bad Seed</u> by Jory John).
- Scenario cards (prepared in advance) with various situations where students can discuss and make judgments (suggestions follow lesson plan).
- Assessment sheet (sample follows lesson plan).

Procedure

Introduction (5 minutes):

a. Begin by discussing with students what judgment means.
b. Ask students to think of times when they would use their judgment to make decisions, e.g., choosing friends, deciding between healthy snacks or junk food, or resolving a conflict.

Discussion: Good Judgment (10 minutes):

a. Read aloud a picture book that focuses on using good judgment (suggestions above).
b. Pause throughout the reading to discuss key points, e.g., characters' decisions, and negative and positive consequences.
c. Encourage students to share what they would do in those situations.

Activity: Scenario Cards (20 minutes):

a. Divide the class into small groups.
b. Distribute scenario cards to each group.
c. Have students read the scenarios and discuss ways to respond in each scenario that show good judgment.
d. After discussion, each group shares their thoughts with the class.

Conclusion (5 minutes):

a. Brief class discussion recapping using good judgment.

Assessment (10 minutes):

a. Distribute the assessment worksheet (sample provided below).
b. Collect the worksheets for assessment.

Homework (optional):

a. Encourage students to share what they learned about using good judgment with their families.

Suggestions for Scenario Card:

1. Scenario: A classmate is being teased by a group of students. What would you do?
2. Scenario: You find a wallet on the ground with identification and cash in it. What should you do with it?
3. Scenario: Your friend wants you to help them cheat on a test. What would you say to them?
4. Scenario: You accidentally break a classmate's favorite pencil. What is the right thing to do?
5. Scenario: You see someone littering in the park. What should you do?
6. Scenario: Your little brother/sister wants to play with your favorite toy, but you are worried they might break it. How do you handle this?
7. Scenario: You are at a friend's house, and they offer you a snack that you know you are supposed to eat. What do you do?

8. Scenario: While playing a game with friends, one of your friends keeps cheating. What do you do?

9. Scenario: At the library, you see someone tearing pages out of a book. What's the best thing to do in this situation?

10. Scenario: Your teacher accidentally gives you candy that you know is not yours. You know who it belongs to. What would you do?

Modify or add scenarios based on your students' specific needs and experiences.

Sample Assessment Sheet:

Good Judgment Assessment Worksheet:

Name: _____ Date: _____

Directions: Read each situation carefully then decide what you would do. Circle the letter next to the action you think shows good judgment.

1. During lunch, you see a classmate drop their lunchbox, and all their food spills out. What do you do?
   a. Laugh and walk away.
   b. Help them pick up their food and ask if they're okay.
   c. Ignore them and continue eating your lunch.
   d. Push them and tell them it's their fault.
      Answer: b. Help them pick up their food and ask if they're okay

2. You're at a friend's birthday party, and they offer you a slice of cake. You're on a diet and trying to eat healthier. What should you do?
   a. Politely decline the cake and explain why.
   b. Take the cake but throw it away when no one's looking.
   c. Accept the cake and eat it even though you know it's not good for you.
   d. Ignore the cake and go play with your friends.
      Answer: a. Politely decline the cake and explain why

3. Your little sister accidentally spills juice on your favorite book. How do you react?
   a. Yell at her and tell her she's not allowed to touch your things anymore.
   b. Calmly explain that accidents happen and help her clean it up.
   c. Ignore it and go tell your mom to punish her.
   d. Hit your sister because you're angry.
      Answer: b. Calmly explain that accidents happen and help her clean it up

4. You're playing tag with your friends, and one of them falls and gets hurt. What's the right thing to do?
   a. Laugh at them because it's funny.

b. Run away and pretend you didn't see it.

c. Stop playing and check if they're okay, then go get help if needed.

d. Keep playing and ignore them.

Answer: c. Stop playing and check if they're okay, then go get help if needed

5 Your teacher accidentally gives you extra stickers, but they're meant for another student. What would you do?

a. Keep the stickers for yourself and not say anything.

b. Give the stickers to the other student and tell your teacher about the mistake.

c. Hide the stickers so no one finds out.

d. Share the stickers with your friends.

Answer: b. Give the stickers to the other student and tell your teacher about the mistake

Scoring: Each correct answer is worth 1 point. Add up your points to see how well you did:

- 5 points: Excellent! You demonstrated good judgment in all the scenarios.
- 4 points: Very good! You made some very good choices, but there's always room for improvement.
- 3 points or below: Room to grow. Think about what you could do differently in each scenario.

  Feel free to adjust the scenarios and answer choices based on your students' needs.

TOC Title: G3 ChatGPT
  Lesson Title: ChatGPT
  Grade Level: 3
  Duration: 35 minutes (without mini project which may be
completed as a separate lesson plan)

Objective:

- The students will show a basic understanding of ChatGPT and AI assistants.
- The students will be able to think critically about technology and its potential impact.

Suggested Materials:

- Whiteboard and markers.
- Projector or screen for displaying examples.
- Internet access (for demonstration if available).

Procedure

Introduction (5 minutes):

a. Ask students if they know what artificial intelligence (AI) is, and if they have heard of AI assistants like Siri or Alexa.
b. Full group: Introduce the topic of Chat GPT and AI assistants.

What is ChatGPT? (10 minutes):

a. Explain that ChatGPT is a kind of AI designed to understand and create text responses that sound human.
b. Describe how ChatGPT works: it learns from a large amount of text data and uses patterns to generate responses.
c. Provide examples of how ChatGPT can be used, e.g., answering questions, generating stories, or providing recommendations. These should be projected for the class to see.

Ethical Considerations (5 minutes):

a. Discuss with the students the importance of using technology responsibly and ethically.
b. Talk about privacy concerns related to AI assistants and the importance of respecting others' privacy when using technology.

Reflection and Discussion (5 minutes):

a. Ask students what they think about ChatGPT and AI technology in general.
b. Discuss potential benefits and drawbacks of AI conversational assistants.

Extension Activity — optional:

ChatGPT Mini-Project:

a. Students can write a plan for their own AI assistant and present it to the class, explaining its features and functions.

Conclusion (5 minutes):

a. Summarize the key points covered in the lesson.
b. Encourage critical thinking about technology and the impact it can have.

Assessment (informal):

a. Observe students' engagement and participation in discussions.
b. Check students' understanding of key concepts, e.g., how ChatGPT learns and its potential applications.
c. Check students' comprehension of the ethical considerations of AI technology.

Grade 4

TOC Title: G4 Acceptable Behavior
  Lesson Title: Acceptable Behavior
  Grade Level: 4
  Duration: 30 Minutes (not including assessment)

Objective:

- The students will understand and demonstrate acceptable behavior in given scenarios.

Suggested Materials:

- Whiteboard and markers.
- Chart paper.
- Sticky notes.
- Scenarios cards (sample after lesson plan).

Procedure

Introduction (5 minutes):

a. Begin with a brief discussion about what is meant by acceptable behavior.
b. Ask students to share what they consider examples of acceptable behavior.

Activity 1: Acceptable Behavior in the Classroom (15 minutes):

a. With input and consensus from the students, write out a list of classroom behavior expectations on the whiteboard, e.g., raising your hand, listening to others, following directions, being respectful toward everyone, etc.
b. Discuss the expectations with the students. Provide both good and bad examples of behavior for each.
c. Put the class into small groups (3–4 students). Provide each group with a scenario card (sample follows lesson plan).
d. Each group will discuss their scenario and come up with appropriate behavior based on the classroom expectations.
e. Have each group present their scenario and discuss the acceptable behavior with the whole class.

Conclusion (10 minutes):

a. Whole group: review the key concepts covered about acceptable behavior.

b. Facilitate a class discussion on how everyone can apply what they have learned in different situations at school, at home, and in the community.

Assessment:

a. An informal assessment can be made based on each student's participation in class discussions, their ability to identify and demonstrate acceptable behavior in given scenarios, and their responses during the conclusion.

Acceptable Behavior Sample Scenarios: (feel free to replace, change, or add to any examples)

1. During group work, one of your classmates keeps interrupting others while they are speaking. What should you do?
2. You notice a student sitting alone at lunchtime. What could you do to include them?
3. Your friend is struggling with a math problem. How can you help them without doing the work for them?
4. Your teacher asks you to clean up your desk, but you're in the middle of reading an interesting book. What should you do?
5. You accidentally bump into someone in the hallway. What is the appropriate response?
6. Your friend wants to copy your homework. What should you say or do?
7. You see someone being bullied on the playground. What actions could you take to help them?
8. Your friend is sad because they didn't do well on a test. How can you support them?
9. Your teacher assigns group work, and you disagree with your group's idea. How can you express your opinion respectfully?
10. You finish your work early and notice a classmate struggling to finish theirs. What could you do to help them?

TOC Title: G4 Cyberbullying
  Lesson Title: Cyberbullying
  Grade Level: 4
  Duration: 50 minutes

Objectives:

- Students will be able to define cyberbullying and comprehend its impact on individuals.
- Students will be able to identify different forms of cyberbullying.
- Students will be able to use strategies to prevent and respond to cyberbullying.

Suggested Materials:

- Whiteboard and markers.
- Large paper or poster board, markers, and/or crayons.
- "NetSafe Episode 5: Cyberbullies are No Fun!" (https://www.youtube.com/watch?v=peDosNN7l3w) or another video that addresses the topic at an age-appropriate level.

Procedure

Introduction (10 minutes):

a. Ask students if they know what the term "cyberbullying" means. Write their responses on the whiteboard.
b. Give the definition of cyberbullying: the use of technology, such as the Internet or social media, to harass, intimidate, or harm others.
c. Discuss the importance of being kind online and treating others with respect, just like in real life outside of the digital world.

Video and Discussion (15 minutes):

Discussion:

a. Show age-appropriate video (option suggested above) on Internet safety and cyberbullying. Discuss the key points with the class.

b. Create a chart on the whiteboard listing different forms of cyberbullying (e.g., mean comments, spreading rumors, exclusion, and impersonation).

c. Lead a class discussion on how cyberbullying might make someone feel and why it is important to prevent it.

Activity (20 minutes):

Digital Kindness Poster Creation:

a. Divide the class into small groups (2–4 students or as appropriate by class size).

b. Hand out art supplies.

c. Ask students to design posters promoting digital kindness and anti-cyberbullying messages. They can use drawings, slogans, and key points discussed in class.

d. Display the posters around the classroom or school to raise awareness.

Conclusion (5 minutes):

a. Have students to reflect on what they have learned about cyberbullying.

b. Discuss ways everyone can promote a positive online environment.

Assessment:

a. Evaluate students' participation in a small-group discussion environment and how their posters incorporate the message of the lesson.

Homework (optional):

a. Encourage students to share what they learned about cyber safety with their families.

TOC Title: G4 Physical Bullying
Lesson Title: Physical Bullying
Grade Level: 4
Duration: 45 minutes

Objective:

- Students will understand the concept of physical bullying and its impact.
- Students will recognize different forms of physical bullying.
- Students will learn strategies to prevent and respond to physical bullying.

Suggested Materials:

- Whiteboard and markers.
- Chart paper and markers.
- Video on physical bullying (Suggestion: "What Is Bullying? for Kids | How to Stop Bullying" | National Bullying Prevention Month | Twinkl USA (https://www.youtube.com/watch?v=ffzIhWoi5ac) or another video that addresses the topic at an age-appropriate level).
- Worksheets or handouts on physical bullying.

Procedure

Introduction (15 minutes):

a. Begin with a general discussion about bullying. Ask students to share what they think bullying is.
b. Define physical bullying, emphasizing that it involves using physical force to harm or intimidate others.
c. Discuss the different forms physical bullying can take (e.g., hitting, kicking, and pushing) and emphasize that it is never okay.

Understand Physical Bullying (15 minutes):

a. Use the whiteboard to list examples of physical bullying. Encourage students to share their own experiences without disclosing personal details.

b. Show anti-bullying video (one possible suggestion above) to help students visualize the impact of physical bullying on individuals and the school community.

c. Discuss the consequences of physical bullying, both for the victim and the perpetrator.

Preventing Physical Bullying (10 minutes):

Introduce strategies for preventing physical bullying. These may include:

a. Developing empathy: Discuss the importance of understanding and respecting others' feelings.

b. Reporting incidents: Encourage students to report any instances of physical bullying to a trusted adult.

c. Conflict resolution: Teach basic conflict resolution skills, emphasizing the importance of communication.

d. Building a positive environment: Discuss ways students can contribute to creating a positive and inclusive school environment.

Conclusion (5 minutes):

a. Summarize the key points discussed during the lesson.

b. Reiterate the importance of working together to create a safe and respectful school environment.

Assessment:

a. Make an informal assessment of student learning during discussions regarding types of bullying and strategies to prevent physical bullying.

Extension:

a. Revisit these concepts frequently to reinforce what the students have learned.

TOC Title: G4 Verbal Bullying
Lesson Title: Verbal Bullying
Grade Level: 4
Duration: 45 minutes

Objective:

- The students will be able to define and identify verbal bullying.
- The students will help develop strategies to respond to verbal bullying.

Suggested Materials:

- Whiteboard and markers.
- Chart paper.
- Markers, crayons, or colored pencils.

Procedure

Introduction (10 minutes):

a. Ask students what they know about bullying.
b. Give definition of verbal bullying — using words to hurt, threaten, or tease someone repeatedly.
c. Discuss how verbal bullying can be harmful, and how impacts someone's feelings.

Discussion (10 minutes):

a. Lead a class discussion on different types of verbal bullying. Write examples on the whiteboard. Examples may include name calling, mocking, slurs, humiliating or threatening someone, hurtful teasing, and racist comments.
b. Ask students to share any experiences they may have experienced or witnessed with verbal bullying. Ensure a safe environment.

Activity (20 minutes):

Class Chart:

a. Create a "Positive Communication" chart on large paper.
b. Students suggest positive and encouraging words or phrases that can be used instead of hurtful ones.
c. Let students use colored markers to decorate the chart.

Conclusion (5 minutes):

a. Ask students how they can help prevent verbal bullying and create a positive and respectful classroom environment.
b. Ask students to share one thing they learned today about verbal bullying.

Assessment:

a. An Informal assessment of student participation through observation during class role-playing activities and discussions.

Extension Activity (optional):

a. Encourage students to create posters with anti-bullying messages, including kind, supportive words and phrases.
b. Display the posters around the classroom or school.

TOC Title: G4 Emotional Bullying
  Lesson Title: Emotional Bullying
  Grade Level: 4
  Duration: 50 minutes

Objective:

- The students will be able to identify emotional bullying.
- The students will be able to use strategies to respond to emotional bullying.

Suggested Materials:

- Whiteboard and markers.
- Storybooks depicting emotional bullying. Suggestion: *Just Kidding* by Trudy Ludwig or another book that addresses the topic at an age-appropriate level.
- Index cards.
- Art supplies (markers, colored pencils, paper).

Procedure

Introduction (10 minutes):

a. Start by discussing why it is important to treat others with kindness and respect.
b. Give the definition of emotional bullying as repeatedly hurting someone's feelings or manipulating their emotions on purpose.

Read a picture book that deals with the topic. *Just Kidding* by Trudy Ludwig is suggested above (10 minutes).

Activity (20 minutes):

a. Brainstorm on the whiteboard different forms of emotional bullying, encouraging students to share examples.
b. Differentiate between playful banter versus hurtful comments meant to harm.
c. Divide the class into small groups (2–4 students each).
d. Give each group markers/pencils and index cards.

e. Let each group create a rule or guideline related to preventing emotional bullying. Examples of rules/guidelines include no bullying, treat others with respect, use kind words, etc.

f. Compile the rules on a large chart paper to create a class charter that can be displayed in the classroom.

Conclusion (10 minutes):

a. Students and teachers sign the charter as a commitment to creating a positive, respectful environment.

b. Remind students that it is important to report emotional bullying to teachers or trusted adults whenever it is experienced/witnessed.

Assessment:

a. Make an informal assessment of student participation and contribution during the development of the rules/guidelines for the charter development.

TOC Title: G4 Social Bullying
Lesson Title: Social Bullying
Grade Level: 4
Duration: 40 minutes

Objective:

- Students will be able to define social bullying and will understand its impact on individuals.
- Students will develop strategies to prevent and respond to social bullying.

Suggested Materials:

- Whiteboard and markers.
- Videos on empathy and kindness such as "All about Empathy (For Kids!)" (https://www.youtube.com/watch?v=Itp21tly8nM) (youtube.com) or another video that addresses the topic at an age-appropriate level.
- Small blank cards for students.

Procedure

Introduction (10 minutes):

a. Begin with a class discussion on bullying. Ask students to share their descriptions. Write their responses on the whiteboard.
b. Introduce the term social bullying. Explain that it involves intentionally excluding, gossiping, or spreading rumors about someone, trying to damage someone's reputation.
c. Discuss the emotional impact of social bullying on individuals and emphasize the importance of creating a positive and inclusive environment.

Video and Discussion (10 minutes):

a. Show an age-appropriate educational video (suggestion above) that highlights the importance of empathy, kindness, and inclusion.
b. Discuss important topics from the video.

Activity (20 minutes):

Cyber Safety Poster Creation

a. Distribute small blank cards to students.
b. Students write or draw positive messages or actions that can help prevent social bullying.
c. Collect cards and create a classroom display to serve as a reminder of the students' commitment to promoting kindness and inclusion.

Conclusion (5 minutes):

a. Review the key points from the lesson.
b. Emphasize the importance of being proactive in creating and maintaining a supportive, respectful classroom environment.

Assessment:

a. Evaluate students' participation in discussions and how their drawing activity connects with the message of the lesson.

Homework (Optional):

a. Ask students to talk about kindness with their families.

TOC Title: G4 Disability Bullying
  Lesson Title: Disability Bullying
  Grade Level: 4
  Duration: 50 minutes

Objective:

- The students will understand what disability bullying is.
- The students will recognize the impact of disability bullying on individuals.
- The students will learn strategies to prevent and address disability bullying.

Suggested Materials:

- Whiteboard and markers or a projector for visuals.
- Picture books or short videos about disabilities and inclusion. Suggestion: "Children with Special Needs" from Jason I Am (https://www.youtube.com/watch?v=5uVoyyNJKys) or another video that addresses the topic at an age-appropriate level.
- Chart paper and markers.

Procedure

Introduction (15 minutes):

a. Lead a whole group discussion on bullying. Ask students if they are aware of different types of bullying.
b. Introduce the concept of disability bullying, explaining that it involves mistreating someone because of their disability.
c. Use real-life examples or scenarios to illustrate disability bullying.
d. Talk about the impact disability bullying has on individuals. Emphasize the emotional and psychological cost to the victims.

Discussion (15 minutes):

a. Show video (one is suggested above).
b. Discuss the importance of empathy and understanding.

Activity (15 minutes):

Preventing Disability Bullying:

a. Discuss strategies to prevent disability bullying, such as promoting inclusion, educating others about disabilities, and fostering a supportive environment.
b. Create a class chart together, listing ways to prevent disability bullying.
c. Display chart in the classroom.

Conclusion (5 minutes):

a. Highlight the importance of treating people with respect and kindness regardless of their differences.

Assessment:

a. Observe how the students summarize the main points of the lesson during the discussion.

TOC Title: G4 Good Judgment
   Lesson Title: Good Judgment
   Grade Level: 4
   Duration: 45 minutes (not including assessment)

Objective:

- Students will know what is meant by good judgment.
- Students will be able to use strategies for using good judgment in various situations.

Suggested Materials:

- Whiteboard or chalkboard.
- Markers or chalk.
- Chart paper.
- Sticky notes.
- Pencils and paper.
- Assessment worksheet (suggestion below).

Procedure

Introduction (10 minutes):

a. Discuss with the students what they think "good judgment" means. Encourage them to share their ideas and give examples.
b. Write down key points on the board to show an agreed upon definition of good judgment.

Activity 1: Good Judgment Strategies (15 minutes):

a. Create a list of strategies students can use to make good decisions. Include ideas such as: thinking before doing, considering consequences, asking trusted adults for advice, considering the feelings of others, using past experiences, etc. The list should be posted for students to see, i.e. on large chart paper or the whiteboard.
b. Discuss each strategy with the students.

Activity 2: Using Good Judgment (15 minutes):

   a. Give a sticky note to each student.
   b. Ask students to write down one example of a time when they used good judgment.
   c. Have the students put their sticky notes on the board.
   d. Read some of the examples as a class and discuss. Highlight the different ways good judgment can be demonstrated.

Conclusion (5 minutes):

   a. Recap the key points of the lesson, emphasizing the importance of good judgment.

Sample Assessment Sheet:

Good Judgment Assessment Worksheet:

Name: _____ Date: _____

Hand out assessment worksheet (sample below). Students work independently to write responses that demonstrate their understanding of good judgment.

Scenario 1:

You are playing with your friends at the park. One of your friends suggests climbing a tree, but it looks dangerous. What do you do?

Scenario 2:

During a test, you finish early and notice that your neighbor is struggling. They keep looking at your paper. What do you do?

Scenario 3:

Your teacher assigns a group project, and you are paired with someone you do not get along with. How do you handle the situation?

Scenario 4:

You accidentally break a friend's favorite toy. What is the best way to handle this situation?

Scenario 5:

You see someone being bullied on the playground. What should you do?

TOC Title: G4
  Lesson Title: ChatGPT
  Grade Level: 4
  Duration: 55–65 minutes (Consider breaking into two lessons stopping lesson 1 after the "Understanding" section.)

Objective:

- The students will know the basics of artificial intelligence and how ChatGPT functions.
- Students will understand the potential impact of artificial intelligence on society.

Suggested Materials:

- Computers/tablets with Internet access.
- Whiteboard or chart paper.
- Markers.

Procedure

Introduction (5 minutes):

a. Begin with a whole group discussion to assess what the students already know about ChatGPT. Encourage them to share what they know on the topic.
b. Define artificial intelligence (AI) in fourth grade terms, e.g., "Artificial intelligence is when computers or machines are able to perform tasks that would otherwise require human intelligence.

Understanding AI (10 minutes):

a. Use the whiteboard or chart paper to illustrate some examples of AI in everyday life, e.g., virtual assistants (like Siri or Alexa), recommendation systems (like those used by streaming platforms), and self-driving cars.
b. Discuss how AI technologies like ChatGPT work by processing large amounts of data and learning from patterns to generate responses.

   NOTE: Break here if doing two lessons.

Exploring ChatGPT (20 minutes):

a. Divide the students into small groups or pairs. This can be done individually if there is a 1:1 student to device ratio.

b. Provide each group with a computer or tablet and direct them to a website where they can interact with ChatGPT or provide a pre-selected prompt for ChatGPT to respond to. Two suggestions are ChatGPT (https://chat.openai.com/,), and Vello.ai (https://vello.ai/room/yhyex6gj). Other options can be found by searching for free ChatGPT sites. The teacher should check sites before recommending or sending links to students.

c. Encourage students to ask questions, give commands, or engage in a conversation with ChatGPT.

d. Observe the groups' interactions with ChatGPT and offer assistance as needed.

Reflection and Discussion (15 minutes):

a. Bring the class back together for a discussion.

b. Ask students to share their experiences interacting with ChatGPT. What did they find surprising or interesting? Did they find any limitations or challenges in their interactions?

c. Facilitate a discussion on the potential benefits and potential problems of AI technologies like ChatGPT. Encourage students to consider how AI might impact their lives in the future.

d. Emphasize the importance of understanding AI and its implications as they grow up in a world increasingly influenced by technology.

Conclusion (5 minutes):

a. Summarize the key points of the lesson and address any lingering questions or concerns.

Assessment (10 minutes):

a. Ask students to write or type a short reflection on their experience with ChatGPT and what they learned from the lesson.

b. The students should consider the following questions in their response:

- What did you find most interesting or surprising about ChatGPT?
- Did you encounter any challenges or limitations while interacting with ChatGPT?
- If so, what were they?
- How do you think AI technologies like ChatGPT could be useful in the future?
- Are there any concerns or questions you have about AI after participating in this lesson?

Notes

1 Betkowski, A., August 01, 2023, "Teaching Tuesday: What Is Character Education?" Teaching & School Administration, Grand Canyon University, www.gcu.edu/blog/teaching-school-administration/what-character-education#:~:text=Character%20education%20aims%20to%20 develop,a%20world%20worth%20living%20in, retrieved January 29, 2024.

2 OpenAI, "How Do We Define Character Education?" GPT-3.5, https:// chat.openai.com/c/e15869b4–2055-44f6-a69e-76aea0a13b40, prompt submitted November 12, 2023.

3 (n.a.), 2023, "Essential Components of MTSS," American Institutes for Research, https://mtss4success.org/essential-components#:~:text=A%20 multi%2Dtiered%20system%20of,from%20a%20 strengths%2Dbased%20perspective, retrieved January 29, 2024.

4 (n.a.), 2023, "What is PBIS?," Center on PBIS, Positive Behavioral Interventions & Supports, www.pbis.org/pbis/what-is-pbis, retrieved November 10, 2023.

5 (n.a.), 2021, "RTI vs. MTSS," Interval Technology Partners, LLC, www.enrichingstudents.com/rti-vs-mtss/#:~:text=RTI%20is%20con-sidered%20a%20more,%2C%20and%20social-emotional%20support, retrieved January 29, 2024.

6 (n.a.), (n.d.), "11 Principles in Schools," Character.org (formerly the Character Education Partnership), https://character.org/11-princi-ples-in-schools/, retrieved January 29, 2024.

7 Shutterstock Image, February 11, 2024, used under CRC License.

8 (n.a.), March 2002, "U.S. Department of Education Strategic Plan, 2002–2007," Office of the Deputy Secretary (ED), Washington, DC, Pg. 6, and Pg 16, https://www.govinfo.gov/content/pkg/ERIC-ED466025/ pdf/ERIC-ED466025.pdf, retrieved January 31, 2024.

9 (n.a.), (n.d.), "Character Education…Our Shared Responsibility," U.S. Department of Education, Office of Safe and Drug-Free Schools, www2.ed.gov/admins/lead/character/brochure.html, retrieved January 31, 2024.

10 Sugiarti, R., Erlangga, E., Suhariadi, F., Winta, M. V. I., & Pribadi, A. S., April 27, 2002, "The Influence of Parenting on Building Character in Adolescents," *Heliyon*, 8(5), e09349. https://doi.org/10.1016/j.heli-yon.2022.e09349. PMID: 35586332; PMCID: PMC9108886. Open access article under the CC BY-NC-ND license (http://creativecom-mons.org/licenses/by-nc-nd/4.0/), retrieved January 31, 2024.

11 The Foundation of Character- Core Values, The graphic in Figure 1.2 was designed and produced by the authors.

12 The 14 core values were compiled and summarized from multiple sources including (1) Carpenter, M., March 11, 2022, "What are core values, and how do you pick them for your characters?" https://medium.com/ keyboard-quill/what-are-core-values-and-how-do-you-pick-them-f or-your-characters-aa3b85cf3b68#:~:text=A%20core%20value%20

is%20a,guided%20by%20their%20core%20values; (2) (n.a.), (n.d.), "Character Education…Our Shared Responsibility," U.S. Department of Education, Office of Safe and Drug-Free Schools, www2.ed.gov/admins/lead/character/brochure.html; (3) (n.a.), (n.d.), "Five Core Values," The National Association of Intercollegiate Athletics (NAIA), www.naia.org/champions-of-character/five-core-values; (4) (n.a.), (n.d.), "The Six Pillars of Character," Character Counts!, https://character-counts.org/six-pillars-of-character; (5) OpenAI at www.openai.com; (6) Spallino. J., January 23, 2017, "How Character Education Helps Kids Learn and Develop," Service Learning, www.methodschools.org/blog/how-character-education-helps-kids-learn-and-develop, and (7) Sugiarti, R., Erlangga, E., Suhariadi, F., Winta, M. V. I, & Pribadi, A. S., April 27, 2002, "The Influence Of Parenting On Building Character In Adolescents," *Heliyon*, 8(5), e09349. https://doi.org/10.1016/j.heliyon.2022.e09349. PMID: 35586332; PMCID: PMC9108886, retrieved January 31, 2024.

13 Kittelstad, K., n.d., "What's the Difference Between Ethics, Morals and Values?" https://examples.yourdictionary.com/difference-between-ethics-morals-and-values.html, retrieved February 1, 2024.

14 OpenAI, "Provide a brief definition of norms as used in the field and study of ethics," GPT-3.5, OpenAI's large-scale language-generation model, https://chat.openai.com, chat.openai.com/c/6c151798-9d2d-482b-a075-db136c385112, prompt submitted November 18, 2023.

15 The Core Principles of Ethics, Figure 1.7 was designed and produced by the authors.

16 Velasquez, M., Moberg, D., Meyer, M., Shanks, T., et al., November 5, 2021, "A Framework for Thinking Ethically," The Markkula Center for Applied Ethics at Santa Clara University, www.scu.edu/ethics/practicing/decision/framework.html, retrieved February 10, 2024.

17 The Anti-Bullying Alliance. https://anti-bullyingalliance.org.uk/aba-our-work. Retrieved January 18, 2024.

18 (n.a.) "Fast Facts: Preventing Bullying", The Centers for Disease Control and Prevention, www.cdc.gov/violenceprevention/youthviolence/bullyingresearch/fastfact.html#:~:text=Bullying%20is%20a%20frequent%20 discipline,and%20primary%20schools%20(9%25), retrieved January 18, 2024.

19 StopBullying.gov. www.stopbullying.gov/resources/laws, retrieved January 18, 2024.

20 University of the People. "Definition of Bullying," www.uopeople.edu/blog/definition-of-bullying/, retrieved January 22, 2024.

21 Preventing and Promoting Relationships & Eliminating Violence Network. "Types of Bullying," www.prevnet.ca/bullying/types, retrieved January 22, 2024.

22 (n.a.) "Fast Facts: Preventing Bullying", The Centers for Disease Control and Prevention, https://www.cdc.gov/violenceprevention/youthviolence/bullyingresearch/fastfact.html#:~:text=Bullying%20is%20a%20frequent%20discipline,and%20primary%20schools%20(9%25), retrieved January 24, 2024.

23 Shutterstock image, March 9, 2024, used under CRC License.

24 (na.), (n.d.), www.unesco.org/en/days/against-school-violence-and-bullying, retrieved January 25, 2024

25 StopBullying.gov, date last reviewed: October 6, 2021. "Federal Laws." www.stopbullying.gov/resources/laws/federal, retrieved February 8, 2024.

26 StopBullying.gov, date last reviewed: May 17, 2023. "Laws and Policies." www.stopbullying.gov/resources/laws/, retrieved February 8, 2024.

27 Shutterstock image, March 9, 2024, used under CRC License.

28 Lee, A. M. I., 2024, "The Difference Between Teasing and Bullying", www.understood.org/en/articles/difference-between-teasing-and-bullying, retrieved January 25, 2024.

29 Johnson, N., and Stixrud, W., 2024, Psychology Today. Posted August 18, 2022. "Helping Kids Become Good Decision Makers." www.psychologytoday.com/us/blog/the-self-driven-child/202208/helping-kids-become-good-decision-makers, retrieved January 31. 2024.

30 (n.a.), "Child's Play: Kids as Young as Six Consider Choices in Moral Judgments", Neuroscience News, https://neurosciencenews.com/moral-judgment-child-23350/, retrieved February 1, 2024.

31 Miller, Gia, Child Mind Institute. October 23, 2023. *Helping Kids Make Decision*, retrieved February 15, 2024.

32 Shutterstock image, February 29, 2024, used under CRC License

33 (n.a.), "Tips for Helping Children Develop Healthy Decision-making Habits", Wellspring Center for Prevention, https://wellspringprevention.org/blog/help-child-develop-decision-making-skills, retrieved February 9, 2024.

34 (n.a.), "Child's Play: Kids as Young as Six Consider Choices in Moral Judgments", Neuroscience News, https://neurosciencenews.com/moral-judgment-child-23350, retrieved February 9, 2024.

35 (n.a.), "Why Do Some Kids Take Bigger Risks Than Others?", Neuroscience News, Why Do Some Kids Take Bigger Risks Than Others? - Neuroscience News, retrieved February 15, 2024.

36 Defining Technology for 3rd and 4th Grade Students, Blackboard Photo and Text designed by authors. March 2024.

37 (n.a.), January 21, 2020, "Trustworthy AI: A Q&A With NIST's Chuck Romine," www.nist.gov/blogs/taking-measure/trustworthy-ai-qa-nists-chuck-romine, retrieved February 1, 2024.

38 Cotton, P., Patel, M., & Wei, W., May 2022, The foundational standards for AI ISO/IEC 22989 and ISO/IEC 23053, ISO/IEC AI Workshop, https://jtc1info.org/wp-content/uploads/2022/06/03_08_Paul_Milan_Wei_The-foundational-standards-for-AI-20220525-ww-mp.pdf, retrieved February 11, 2024.

39 Shutterstock image, February 11, 2024, used under CRC License.

40 Marcella, A., November 16, 2023, "Please explain what generative AI is using non-technical terms," ChatGPT 4, OpenAI, https://chat. openai.com/share/cc8db088-0c4f-4426-a3d2-b4763ff82e12, retrieved February 11, 2024.

41 (n.a.), (n.d.), What are AI hallucinations? Google Cloud, https:// cloud.google.com/discover/what-are-ai-hallucinations#:~:text=st arted%20for%20free-,How%20do%20AI%20hallucinations%20 occur%3F,making%20incorrect%20predictions%2C%20or%20halluci-nating, retrieved February 10, 2024.

42 Barney, N., March 2023, "deepfake AI (deep fake)," TechTarget, www. techtarget.com/whatis/definition/deepfake, retrieved February 10, 2024.

43 Significant Issues of Unethical AI, February 11, 2024, graphic drawn by authors.

44 Barry, P., October 12, 2023, "The Meaning Behind the Song: Teach Your Children by Crosby – Stills – Nash & Young," https://old-timemusic.com/the-meaning-behind-the-song-teach-your-children-by-crosby-stills-nash-young, retrieved February 3, 2024.

45 (n.a.), 2023, What is a chatbot? Oracle, www.oracle.com/chatbots/ what-is-a-chatbot, retrieved February 10, 2024.

46 Shutterstock image, February 11, 2024, used under CRC License.

47 Martin, N., June 5, 2023, "The Boom of AI Tools," Nerd For Tech, https://medium.com/nerd-for-tech/the-boom-of-ai-tools-d430060f 5fe0#:~:text=Last%20update%3A%20November%2022th%2C%20 2023,the%20best%20strategy%20with%20them, retrieved February 10, 2024.

48 McGill, J., June 27, 2023, "How Many AI Tools Are There?" Content of Scale, https://contentatscale.ai/blog/how-many-ai-tools-are-there/, retrieved February 10, 2024.

49 Lutkevich, B., January 31, 2024, "34 AI content generators to explore in 2024," TechTarget, https://www.techtarget.com/whatis/feature/ AI-content-generators-to-explore?utm_campaign=20240206_ ERU-ACTIVE_WITHIN_90_DAYS&utm_medium=email&utm_ source=SGERU&source_ad_id=365530155&src=15006417&asrc=EM_ SGERU_287913277, retrieved February 11, 2024.

2

NURTURING DIGITAL CITIZENS

Cyber Safety for Elementary-Aged Children

Technology brings many benefits to users who have learned to use technology responsibly. Awareness, care, caution, and prudent use are the guiding principles when approaching and using technology.

The reality is that technology (in its many forms) also presents potential risks to users when used carelessly, or without taking the proper precautions and safeguards. Users who fail to act responsibly or use caution when embracing technology, whether for business, use in the classroom, or simply for recreational activities may find themselves open to unwanted and oftentimes unseen risks.

As technology grows more pervasive, learning to be cyber-safe when using technology online or offline is an essential skill that, when taught at an early age, will be carried forward throughout one's lifetime.

Introduction

This chapter focuses on online cyber safety. We will discuss the concept and necessity of learning and using safe cyber actions, and how to communicate to students what it means to be cyber-safe and to develop and practice cyber-safe behavior when engaging with technology whether it is for academic studies or recreation activities.

Presented in this chapter is a broad examination of the cyber risks that users, including students, face when engaging with society's increasing dependence on technology. Here we use the very broad term "users," which we intentionally and generically define as adults and children.

                    DOI: 10.1201/9781003466338-2

The primary focus of this chapter is on fundamental cyber safety concepts, risks specific to elementary-aged children, and practical recommendations for educators working with third through fourth-grade students. The emphasis is on age-appropriate strategies, collaboration with parents, and real-life examples to enhance the applicability and impact of cyber safety education in early childhood education settings.

The objective is to provide students with the knowledge and confidence to embrace technology, understand technology, and use technology, while doing so responsibly, safely, and prudently.

Examples via class lesson plans and exercises are provided at the chapter's conclusion to assist the educator in discussing these broader concepts of cyber safety with students.

The Digital Landscape for Elementary-Aged Children

The Pervasiveness of Technology in Education

> Digital technology is transforming the world of work. To produce the knowledge workers of tomorrow, and to maximize the ability of children to learn, it must also be allowed to transform the world of education.[1]

Technology has become deeply integrated across all aspects of modern education, transforming how instruction is delivered and how students learn. Seamless blending of physical and virtual interactions allows learning to transcend barriers and creates digitally augmented, blended instruction models. This represents a fundamental shift as extensive technological immersion makes education the prime territory for deploying pervasive computing.

Easy access to vast information and data-driven insights certainly confers valuable advantages, as do efficiencies in communication and administrative tasks. However, uncontrolled overuse risks jeopardizing accrued intergenerational wisdom and shared humanity. Keeping developmental needs and learner goals at the center can help overcome such pitfalls. Moderation and balance thus remain key guiding principles moving forward.

Ubiquitous access to devices, platforms, and tools enhances interactivity, expands access to information, and enables customized learning experiences catering to diverse needs. Students today have opportunities to learn collaboratively across geographies, receive personalized guidance, and develop critical future-ready skills. Technology facilitates student-centric active learning and develops essential 21st-century competencies like digital literacy.

Integrating technology throughout instruction equips students with relevant capabilities and prepares them for technology-driven careers. It streamlines administrative processes and provides professional development opportunities for educators as well. Technology-enabled individual-centric customizations are vital for relevance and success in education.

However, while promising, there are also emerging concerns surrounding equitable access, ethical use, and respecting human connections in learning. Addressing these responsibly while harnessing technology's potential requires mindful, balanced application focused on pedagogical objectives over capabilities alone. Ongoing teacher training and investments to bridge digital divides are equally necessary.

Still, suitably incorporated, educational technology can enrich learning and make it enjoyable, engaging, and effective. By interweaving physical and virtual interactions, learning can transcend spatial, chronological, and social barriers. With sound strategies on fronts like teacher readiness and equitable access, integrated ed tech promises to make quality, personalized learning truly scalable and sustainable.

The path forward warrants continued efforts to enable disadvantaged groups through access mechanisms, alongside policy measures around healthy usage norms and privacy considerations amidst rapid technological shifts.

Success lies not in technology itself but in how solutions are judiciously designed, aligned, and scaled based on pedagogical priorities — while addressing inequities responsibly. By maintaining a reasonable balance, technology and education can nurture each other potently. Uncontrolled overuse however risks losing accrued wisdom of generations past.

In essence, while transforming 21st-century education fundamentally, appropriate regulation and customization of technology based on developmental needs and learning objectives remain key. Moderation and balance are crucial guiding principles to realize the promise while overcoming the pitfalls on the road ahead.[2]

The Pervasiveness of Technology: Impact on Elementary-Aged Children

The pervasiveness of technology in education as discussed above refers to the widespread integration and influence of technology across various aspects of the educational landscape. It signifies the extensive use of digital tools, devices, and platforms to enhance teaching and learning experiences at all levels of education. This concept acknowledges that technology is no longer confined to a specific subject or classroom but has become an integral and ubiquitous component of the entire education ecosystem.

Artificial intelligence (AI) will only increase the speed, depth, complexity, and benefits of technology's influence and impact in the classroom and on students. Integrating technology throughout the learning experience equips students with digital skills and prepares students to meet the demands of technology-driven careers.

AI will contribute immensely to the academic and personal growth of elementary-age school children as it will facilitate collaborative learning, connect students globally, and support data-driven instruction.

Listed below are contributions that AI brings to the academic classroom. This list is by far not complete and will only expand over time as new AI tools and the application of those tools mature.

AI's contributions to the classroom include but are certainly not limited to:

1. Personalized Learning: AI-powered educational tools can adapt to each child's learning pace and style, providing customized learning experiences tailored to their needs and abilities.
2. Interactive Learning: AI applications can offer interactive simulations, games, and activities that engage young learners and make learning more enjoyable and memorable.
3. Instant Feedback: AI systems can provide immediate feedback on assignments and exercises, allowing children to identify mistakes and learn from them in real time.
4. Enhanced Accessibility: AI technologies can assist children with disabilities by providing alternative modes of learning and communication, ensuring that all students have equal access to education.

5. Augmented Teaching: AI can assist teachers by automating administrative tasks, providing insights into student progress, and offering personalized recommendations for instructional strategies.

6. Encouragement of Critical Thinking: AI-based learning tools can challenge students with problem-solving tasks that promote critical thinking, creativity, and innovation.

7. Expanded Curriculum: AI can supplement traditional curriculum materials with additional resources, including educational videos, interactive tutorials, and digital textbooks, enriching the learning experience.

8. Exposure to Emerging Technologies: Introducing AI at an early age familiarizes children with cutting-edge technologies, preparing them for future careers in STEM fields and fostering digital literacy.

9. Cultivation of Curiosity: AI-powered educational tools can stimulate curiosity by providing access to vast amounts of information and encouraging exploration and discovery.

10. Preparation for the Future: As AI becomes increasingly prevalent in society, exposing children to AI in education equips them with the skills and knowledge needed to navigate and thrive in a technology-driven world.

These benefits collectively contribute to a more engaging, effective, and inclusive learning environment for elementary school children in grades three through four.

With benefits also come risks. Technology is essential both in education and for education. AI will only burrow deeper and deeper into daily in-class student exercises and schoolwork, external assignments, as well as assisting educator and administrative functions. Addressing the inherent risks associated with technology is essential to both protecting our third- and fourth-grade students and preparing them for future roles in society.

The Pervasiveness of Technology: Inherent Risks

The pervasive integration of technology in education has transformed the way we learn and teach, offering many benefits such as better access, interactive learning experiences, and enhanced collaboration. However, this widespread adoption also brings inherent risks that must be considered.

One of the primary concerns is the potential for a widening of the educational divide. As technology becomes more common in classrooms, students of disadvantaged backgrounds may experience a greater lack of access to basic technology and reliable Internet connections, decreasing access to broader educational opportunities. An additional and growing risk is the threat to students' privacy.

The accumulation and retention of vast amounts of digital information about students, from personal information to academic performance indicators, raises concerns about the processing, protection, and long-term retention of these data. Inadequate cybersecurity measures within a school's (or district's) and student's IT system could lead to potential data breaches, leading to a compromise of sensitive student data, which in turn exposes the data owners to various cyber risks (a loss of privacy for example). Technology can expose children to harmful behaviors. The most common of these is cyberbullying.

Excessive screen time and pressure to constantly interact with digital devices can contribute to stress, anxiety, and a decrease in overall well-being. In addition, in some cases, the amount of time a student spends digitally connected to his/her devices raises concerns about digital addiction and its impact on a student's mental health.

During the COVID-19 pandemic, primary education shifted to a digital platform, thus providing a platform for added stressors in the form of increased screen time for schoolchildren.

Attitudes and Experiences in the Online Space

In March 2023, the government-approved regulatory and competition authority for the broadcasting, telecommunications, and postal industries of the United Kingdom, The Office of Communications, commonly known as Ofcom, released a comprehensive overview of children's media experiences in 2022. The report looked at media use, attitudes, and understanding among children aged 3–17.

The Ofcom report disclosed that parents and children identified positive benefits of being online, especially concerning learning (81% children, 84% parents) and building and maintaining friendships (68% children, 65% parents). However, parents of children aged

12–15 (33%) and 16–17 (41%) were more likely than those of younger children (3–11) to disagree with this statement.

Children's actual experiences online were not always positive. Almost three in ten children aged 8–17 (29%) had experienced someone being nasty or hurtful to them via apps or platforms; this contrasted with two in ten having this experience face to face (20%).

Parents expressed concerns about many aspects of children's media use, including being bullied online (70%) or via games (54%), but the most common concerns among parents related to their child seeing content that was inappropriate for their age (75%), or "adult" or sexual content (73%).

Children aged 8–17 were less likely to believe information from social media apps or sites compared to other sources they used: a third said that they believed all or most of what they see on social media to be true (32%), while two-thirds thought the same for news apps and sites (66%), and 77% for websites used for school or homework.[3]

Between the ages of 8–11 children further develop the ability to talk about their thoughts and feelings.[4] Alongside this, they attach emotional importance to having friends, and their relationships with friends may be strong. However, 8–11-year-olds are also more susceptible to peer pressure and the influence of others. Cognitively, they might have an increased attention span and the ability to understand the viewpoints of others.[5]

Children aged 8–11 could be described as developing skills in media. Two-thirds of 8–11-year-olds reported playing games online (67%).

Smartphone ownership shifts markedly in this group, which correlates with the children's transition to secondary school (a school for students intermediate between elementary school and college; usually grades 9–12), and they are likely to be starting to undertake a wider range of online activities without parental intervention.

The Ofcom report provides additional insights on elementary-aged children and their use of technology.

- Children aged 8–11 could be described as developing skills in media. Smartphone ownership shifts markedly in this group, which correlates with the children's transition to secondary school, and they are likely to be starting to undertake a wider range of online activities without parental intervention.

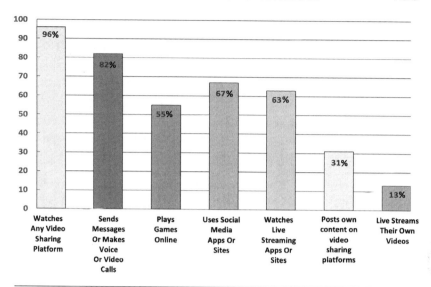

Figure 2.1 Online activities of 8–11-year-olds.[6]

- Children aged 8–11 were doing a wide range of activities online. Most (96%) said they watched videos online, and the majority (63%) also reported using social media platforms. Children aged 8–11 were also more likely than younger children to interact with others by messaging or calling via an app or site, playing video games online, or watching live streams.
- The increase in children who used social media apps at this age was reflected in the types of apps they had profiles on. It was most common for 8–11-year-olds to have profiles on TikTok (32%) and WhatsApp (32%), ahead of YouTube (27%) and Snapchat (24%)[7] (see Figure 2.1).

In a survey of 888 K—12 educators in the U.S. conducted from January 26 to February 7, 2022, by EdWeek Research Centre, 80% of the respondents reported that increased screen time worsened children's behavior.[8]

The rapid pace of technological progress makes it challenging for educators to stay up to date with the evolving tools and platforms. Inadequate training of teachers may obstruct the effective integration of technology into the curriculum and limit its potential benefits. Furthermore, reliance on technology in education can lead to devaluing traditional teaching methods and interpersonal skills.

The proliferation of "Personalized Learning" may lead to an unbalanced delivery of education where silicon-based instructors instead of carbon-based instructors are the primary in-class educators. This may also be cost and budget-driven as school administrators seek to replace the more expensive (salary, benefits, etc.) teachers with less expensive (amortized over many years) technology.

Students may become overly dependent on digital resources, potentially hindering their ability to think critically, solve problems independently, and communicate effectively in face-to-face situations. Additionally, as discussed in Chapter 1, technology (in the form of artificial intelligence (AI) for example) is not always correct.

The need to learn and think independently is a critical skill for learners of all ages. To recap, from Chapter 1, the risks associated with AI that students should be made aware of include:

- Lack of Transparency — Because of the complexity in design and functionality of AI systems it can be difficult for humans to understand how an AI system has come to a particular decision or recommendation.
- Data Privacy — The huge amounts of data used to train AI systems can lead to exposure of private, personal, and sensitive information.
- Vulnerabilities in AI Systems — AI systems can be exposed to attacks just like any other computer system.
- Misuse of AI — AI can be used for malicious purposes, to say mean and untrue things about people. AI can even create completely inaccurate stories.

Furthermore, there is a risk of overemphasis on standardized testing assessment and data-driven metrics in the technology-driven education landscape. The focus on quantifiable results may neglect the development of essential skills such as creativity, critical thinking, and emotional intelligence, which are crucial for success in a rapidly changing world. Lastly, the susceptibility of technology to glitches, outages, and technical issues can disrupt the learning process and create challenges for both educators and students. Reliance on digital platforms without robust backup plans can lead to significant disruptions in the educational environment.

While the integration of technology in education has the potential to revolutionize learning experiences, it is crucial to acknowledge and address the inherent risks associated with its pervasive use. Balancing the benefits of technology with concerns related to accessibility, privacy, mental health, teacher training, skill development, and technical reliability is essential to creating an educational environment that fosters holistic growth and prepares students for the challenges of the future.

Importance of Cyber Safety Education — Why Teach Elementary-Aged Children about Cybersecurity?

Hanover Research's *2023 Trends in K—12 Education* report highlights both new and ongoing issues and priorities that are anticipated to affect K—12 programs in 2023 and presumedly beyond. The report identifies five specific trends that will shape K—12 education. Of specific interest (although all trends identified are of interest) is trend number five (5), "Protecting Student Well-Being Demands Systemic Support."

Hanover states that this trend is "to address the 'whole child' including their social, emotional, mental, and physical needs, schools embrace a systemic approach that targets all elements of student wellness."[9]

Digital technology is rapidly changing the way we educate and protect students in our classrooms. While the Internet can be a tremendous resource for young learners, it also brings a host of potential risks, from cyberbullying to viewing hurtful and sensitive content.

The U.S. Department of Education and the U.S. Department of Health and Human Services identified, in the department's joint report *Policy Brief on Early Learning and Use of Technology*, four guiding principles for the use of technology with early learners.

These four guiding principles are as follows:

- Guiding Principle #1: Technology — when used appropriately — can be a tool for learning.
- Guiding Principle #2: Technology should be used to increase access to learning opportunities for all children.

- Guiding Principle #3: Technology may be used to strengthen relationships among parents, families, early educators, and young children.
- Guiding Principle #4: Technology is more effective for learning when adults and peers interact or co-view with young children.[10]

It is interesting to note some surprising if not concerning information revealed in the *Annual Cybersecurity Attitudes and Behaviors Report 2023* sponsored by the National Cybersecurity Alliance and CybSafe, when asked to respond to the question "I feel that staying secure online is worth the effort" ...

> Of participants responding, 69 percent thought staying secure online was worth the effort. But the younger generations (21% of Gen Z and 23% of Millennials) are skeptical about the return on investment. They were more than twice as likely as Baby Boomers (6%) and the Silent Generation (9%) to doubt online security is worth the effort.[11]

Although technology or the pervasiveness of technology is not specifically addressed in the Hanover Trends report, or the Department of Education and the Department of Health and Human Services report, with the continual evolution of technology and the accelerating use of technology in classrooms and by students, it is imperative that educating children on the responsible use of technology and keeping children safe in the cyber world in which they learn, play, and reside, should be a priority supported by academic administrators and an important objective of K—12 educators.

In essence, technology has transformed teaching and learning by becoming a fundamental, integrated component of education.

Cybersecurity Topics for Elementary-Aged Children

Why are the findings from the Hanover Research's 2023 Trends in K—12 Education report of concern when we examine methods to keep students cyber-safe? If not taught to recognize the risks and to take prudent steps to be cyber-safe while using technology, students could potentially place themselves as well as their family and their friends at risk or in a worse case in danger.

As an example, many web pages are designed to look like legitimate sites that students should trust. As a teaching exercise, educators may wish to address the risks that students may face from this tactic, which include but are not limited to:

- Inappropriate Content Exposure: Fake websites may host inappropriate content that is not suitable for children, including violence, explicit language, or graphic imagery. Children might accidentally stumble upon these sites, leading to potential harm or exposure to content that is not age-appropriate.
- Online Predators: Malicious actors may create fake websites to lure children into sharing personal information or engaging in unsafe online behavior. These predators can pose as friends or trustworthy figures, putting children at risk of exploitation or harm. Read more on this topic in Chapter 3.

When talking about "malicious actors" with elementary-aged children, try this approach:

"Malicious actors" in terms of cyber safety are like sneaky troublemakers who use computers or the Internet to do mean or bad things to others. They might try to trick you into giving them your personal information, like your passwords or where you live, or they might try to damage your computer or make it not work right.

For example, imagine you're playing a game online, and someone you don't know sends you a message asking for your password or your address. They might say they need it to help you win the game or get cool stuff, but really, they just want to do something bad with your information. That person is a malicious actor because they're trying to trick you into giving them something important that could hurt you. It's important to never share personal information online and to tell a grown-up or trusted adult if someone online is acting suspicious or mean.

- Phishing Scams: Fake websites often use deceptive tactics to trick kids into sharing sensitive information, such as passwords or personal details. Children may unknowingly fall victim to phishing scams, compromising their online security.

When explaining "phishing scams" to elementary-aged children, try using the following:

"Phishing scams" are like fishing for information, but on the computer. Just like a fisherman uses bait to catch fish, a phishing scammer uses tricks to try to catch your personal information. They might send you emails or messages that look real like they're from a friend or a company you know, but they're trying to get your secrets, like your passwords or your parents' credit card numbers.

For example, imagine you get an email that says you won a prize, but first, you have to click on a link and enter some information. Even though it looks exciting, it could be a phishing scam. If you click on the link and give them your information, they might use it to do bad things, like take your money or pretend to be you online. It's important to never click on suspicious links or give out personal information online without asking a grown-up or trusted adult first.

- Malware and Viruses: Phony websites can be a breeding ground for malware and viruses. Clicking on links or downloading content from these sites may expose children's devices to harmful software that can damage or compromise their digital devices.

You may wish to try this approach when discussing malware and viruses with elementary-aged children:

"Malware and viruses" are like digital bugs that can make your computer sick. Just like how you can catch a cold from a sneeze, your computer can catch these bugs from the Internet. Malware and viruses can make your computer act funny, slow it down, or even make it stop working altogether.

For example, imagine you're downloading a game from the Internet, but instead of getting the fun game you wanted, your computer starts showing pop-up ads everywhere and won't let you play anything. That could be because you accidentally downloaded malware or a virus along with the game. These digital bugs can sneak onto your computer when you click on the wrong things or visit unsafe websites. That's why it's important to always ask a grown-up or trusted adult before downloading anything from the Internet and to have good antivirus software to protect your computer.

Elementary learners may be unsure or unaware of the term antivirus software. Using the following as an approach to discussing this important safety concept will provide them with the appropriate understanding to keep them cyber-safe.

"Antivirus software" is like a superhero for your computer. It's a special program that helps protect your computer from getting sick with malware and viruses, just like how washing your hands helps protect you from catching germs. Antivirus software works by scanning your computer for any bad bugs and getting rid of them before they can cause any harm.

For example, think of antivirus software as a shield that keeps your computer safe from bad guys. Just like how knights use shields to protect themselves from enemies, antivirus software protects your computer from digital enemies that try to make it sick. It's important to have antivirus software installed on your computer and to keep it updated regularly to stay safe while exploring the Internet.

- Identity Theft: elementary learners may not fully grasp the concept of identity theft, making them vulnerable to fake websites that collect personal information. This can lead to the unauthorized use of their identity for malicious purposes.

To help third and fourth graders better grasp the idea of identity theft and its impact on being cyber-safe the following is a suggested approach:

Teaching Tip

"Identity theft" is like someone pretending to be you, but in the digital world. It happens when someone gets ahold of your personal information, like your name, address, or even your birthday, and uses it to do bad things, like buying stuff or pretending to be you online. Just like how you wouldn't want someone to steal your favorite toy, you definitely don't want someone stealing your identity!

For example, imagine someone gets your name and birthday from your online profile. They could use that information to pretend to be you and buy things with your parent's credit card without permission. Or they could even open a fake social media account pretending to be you and say mean things to your friends. Identity theft can be really scary because it's like someone taking away your online identity and using it for their own purposes.

To protect yourself from identity theft, it's important to never share personal information online unless a grown-up or a trusted adult says it's okay. Also, make sure to use strong passwords and keep them safe. If you think someone might be pretending to be you online or if you see anything suspicious, always tell a trusted adult right away.

- Cyberbullying: Bogus websites may be used as platforms for cyberbullying, where children can be targeted, harassed, or humiliated by others. This can have serious emotional and psychological consequences for the child involved. This topic was addressed in greater detail in Chapter 1.

- Financial Scams: Some fake websites may attempt to trick children into making unauthorized purchases or divulging their parents' financial information. This poses a risk of financial loss and potential legal consequences.

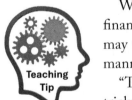

When reviewing the topic of technology-based financial scams with elementary-age learners, you may try approaching the subject in the following manner...

"Technology-based financial scams" are like tricky traps set up by bad people on the Internet to try to take your money or your parents' money. These scams can look real, like a fun game or a cool offer, but they are actually trying to trick you into giving away your money or important information.

For example, imagine you're playing a game online and a pop-up message suddenly appears saying you've won a big prize, but to claim it, you have to pay some money or give them your parents' credit card number. That's a technology-based financial scam! They're trying to trick you into giving them money or personal information, but in reality, there's no prize at all.

These scams can cause a lot of trouble and make people lose their hard-earned money. It's important to always be careful online, never give out personal information or money to strangers, and ask a grown-up or trusted adult if you're not sure if something is safe.

- False Information: Children may encounter fake websites that present false information, which can mislead them in their learning or understanding of various subjects. This misinformation can have educational and cognitive consequences.
- Unsupervised Online Interactions: Phony websites might encourage students to engage in unsupervised online interactions, such as chat rooms or forums, where they could be exposed to inappropriate discussions or interactions with strangers.
- Addiction and Distraction: Fake websites designed to resemble popular games or entertainment platforms may contribute to excessive screen time, leading to potential issues of addiction and distraction from essential activities like homework, physical activities, and family time.

Teaching Tip

When explaining "fake websites" to elementary learners, try this approach:

"Fake websites" are like pretend stores or playgrounds on the Internet. They might look just like the real ones you visit, but they are made by bad people to trick you. These fake websites can try to steal your personal information, like your name or address, or they might even try to get you to buy things that aren't real.

For example, imagine you're looking for a new toy online, and you find a website that looks just like your favorite toy store's website. It has pictures of all the toys you want, but when you try to buy one, it asks for your credit card number and personal information. That could be a fake website! If you give them your information, they might use it to steal your money or your identity.

Fake websites can be tricky, but there are ways to stay safe. Always make sure the website you're on is real by checking the web address and looking for "signs of trust," like reviews or security symbols. And remember, never give out personal information online unless a grown-up says it's okay. If you're not sure if a website is real or fake, it's always best to ask a trusted adult for help.

Some elementary learners may be unaware of "signs of trust" or security symbols. Using the following as an approach to discussing these concepts will provide students with further knowledge on staying cyber-safe when using technology.

"Signs of trust" or security symbols are like special badges that show a website is safe and trustworthy, just like how a superhero's emblem shows they're a hero you can count on. These symbols help you know if a website is real and if it's safe to share your information or buy things from there.

For example, imagine you're on a website looking for a new game to play, and you see a little lock symbol next to the web address at the top of the page. That's a sign of trust! It means the website is secure, like having a secret code that only you and the website know. You can feel safe knowing that your information is protected when you see this symbol.

Other signs of trust could be words like "secure" or "verified" or badges from trusted organizations. These symbols show that the website cares about keeping your information safe and that you can trust them. Remember to always look for these signs of trust before sharing any personal information online or buying anything. If you don't see them, it's best to ask a grown-up or trusted adult for help.

Given the growing level of risk in today's online environment, educators should use the opportunity whenever possible via classroom exercises, and school assignments to communicate the importance of staying cyber-safe to students. Parents and trusted adults can assist educators by providing practical examples and role-playing scenarios to help elementary-age children understand the importance of cybersecurity. See the lesson plans provided in this chapter for some samples and suggestions.

Connected and Protected: Instilling Smart Habits in Young Tech Users

The unavoidable impact of technology on our personal lives is a defining characteristic of our current digital society and the technological environment in which we live and work. From smart home devices that regulate temperature and lighting to virtual assistants that respond to voice commands, technology is seamlessly integrated into almost every aspect of our daily lives.

While technology and continual technological advancements offer convenience and efficiency, they also raise important considerations regarding privacy. Our constant connectivity to devices (mobile phones, tablets, laptops, TVs, etc.) and the data collected for analysis via these devices pose considerable risks to our privacy. Such risks warrant a continual assessment of the importance of balancing the benefits of technological innovation against the loss of personal privacy and an over-dependency on the technology itself.

As our homes become increasingly interconnected, the need to evaluate the ethical use of technology within private spaces (i.e., homes) becomes vital to ensuring a balanced coexistence between digital and personal domains.

Students will need to function, manage, and learn how to remain cyber-safe as omnipresent and all-consuming technology continues to seep into their personal lives and even their own homes. Teaching children about the safety aspects of Smart Home Technology (including the risks and benefits associated with these technologies) is critically important.

Smart home technology involves the integration of devices and systems within a home that can connect to the Internet, enabling users to control and automate various functions remotely. These devices, such as smart thermostats, lights, appliances, entertainment systems, motion sensors, indoor (and exterior) security cameras, and voice-activated assistants, can be managed through a central hub or mobile app. Through this connectivity, users can customize and monitor their home environment, enhancing convenience, energy efficiency, and overall security.

Smart home technology aims to create an interconnected and automated living space that responds intelligently to the preferences and needs of its inhabitants. The aim is to enhance efficiency, convenience, and security by enabling users to remotely manage and automate functions within their homes.

The Internet of Things (IoT) refers to a network of interconnected devices, objects, or "things" embedded with sensors, software, and other technologies, enabling them to collect and exchange data. These devices, ranging from everyday objects like thermostats and appliances to more complex systems like wearable devices, can communicate with each other through the Internet, facilitating data sharing and automation. The goal of IoT is to enhance efficiency, provide new functionalities, and improve overall connectivity in various aspects of daily life.

Interconnected Living: Exploring the Internet of Things in Smart Homes

IoT and Smart Home technology share similarities but differ in scope and application. Both involve the integration of devices with Internet connectivity, allowing for communication and data exchange. However, their primary distinctions lie in their broader purpose and specific focus.

IoT and Smart Home technology involve connecting devices to the Internet for improved functionality, IoT has a broader, industry-spanning scope, whereas Smart Home technology is a subset of IoT, focusing specifically on enhancing residential living spaces. It includes devices like smart thermostats, lights, security cameras, and voice-activated assistants designed to improve convenience and efficiency within homes.

IoT and Smart Home technology introduce specific risks to young learners and children, primarily centered around privacy, security, and potential exposure to inappropriate content. These interconnected devices often collect and share data, without the user's knowledge, raising concerns about the privacy and possible use of youngster's private information. The risk of unauthorized access to personal details and habits poses a threat to the confidentiality of children's data, potentially leading to issues like identity theft or misuse. Additionally, the integration of cameras and microphones in Smart Home devices raises concerns about the unintentional capture of private moments, again impacting the individual's right to privacy.

The risks and security vulnerabilities in IoT and Smart Home technology may expose children to cyber threats, including hacking and phishing attempts. Unauthorized access to connected devices can lead to various risks, such as surveillance, unauthorized control of smart home functions, or the manipulation of devices to expose children to inappropriate content. Furthermore, the increasing dependency on these technologies may contribute to issues like excessive screen time and hinder the development of critical thinking skills as children engage in automated and interconnected environments (read more on this risk and its relationship to cyber-safe practices later in this chapter).

To address these risks, educators must stay informed and discuss with students the need to implement strong privacy settings when engaging with technology and guide children in an overall approach to responsible Internet use. Discussions and class exercises with students can be developed to address the importance of establishing regular communication with a trusted adult. With trusted adults, students can learn to establish appropriate boundaries which contribute to creating a safer digital environment.

The next section examines digital citizenship and continues to reinforce concepts discussed in Chapter 1 on the importance of character development and that good character determines good behavior. A good digital citizen respects others, speaks up when they see something inappropriate, hurtful, or dangerous, and protects themselves and their personal information

Digital Citizenship

Defining Digital Citizenship

Digital citizenship is the ethical, moral, and responsible use of technology to ensure one's own and others' protection while collaborating in an increasingly digital, networked, and global society.

Because of the unique issues that technology has brought into our daily lives, everyone must learn about digital citizenship so that each member of society can become aware of the dangers and pitfalls as well as the positive outcomes associated with taking on the role of the digital citizen in a global community. Although citizenship may be rooted in similar foundational contexts in both offline and online environments, digital citizenship yields many special case issues that must be considered to elicit appropriate and responsible actions in online settings.

For students, the tenants of digital citizenship are far-reaching as the social media technologies with which they are intimately familiar imply several consequences that can have serious implications for their personal, educational, and future business lives.[12]

Expanding this definition to be more inclusive the concept of digital citizenship grows to encompass a range of skills and literacies that can include Internet safety, privacy and security, cyberbullying, online reputation management, communication skills, information literacy, and creative credit and copyright.[13]

What or Who Is a Digital Citizen?

A digital citizen is someone who, through the development of a broad range of competencies, can actively, positively, and responsibly engage in both online and offline communities, whether local, national, or global. As digital technologies are disruptive and constantly evolving, competence building is a lifelong process that should begin from earliest childhood at home and school, in formal, informal, and non-formal educational settings.[14]

Adults possess a keen awareness of the many dangers that children encounter as they navigate through the stages of growth and development. It falls upon the shoulders of these guardians, and in the classroom to educators to equip elementary learners with the necessary skills and knowledge to face these challenges head-on.

Digital citizenship and engagement involve a wide range of activities, from creating, consuming, sharing, playing, and socializing, to investigating, communicating, learning, and working. Digital Competence can be broadly defined as the confident, critical, and creative use of Information and Communication Technologies (ICT) to achieve goals related to work, employability, learning, leisure, inclusion, and/or participation in society.[15]

This responsibility extends beyond the tangible hazards of the physical world to the complex and ever-evolving threats presented by technology (e.g., the Internet). Consequently, in the classroom, educators should be prepared to undertake the comprehensive task of preparing students not only for the realities of life's physical encounters but for virtual interactions and experiences as well. This dual-focused education aims to ensure that elementary-age students are well-versed in navigating both the offline and online aspects of their lives, fostering a safe and informed presence in both domains.

Being a digital citizen involves using technology, especially the Internet, in a manner that respects the rights and well-being of others, while also contributing to the overall betterment of the online community.

Digital citizenship encompasses a range of skills, knowledge, and attitudes that empower individuals to navigate the digital landscape safely and effectively. This includes understanding and practicing concepts such as online etiquette, responsible use of information, digital literacy, and awareness of potential risks and challenges in the digital environment.[16] Becoming a good digital citizen begins with developing, fostering, and sustaining good character and acceptable behaviors, as was discussed in Chapter 1.

A Framework for Digital Citizenship

In 2016, The Council of Europe established the Digital Citizenship Education Expert Group, to investigate good practices in digital citizenship education. Research from this investigation disclosed a lack of awareness among educators of the importance of digital citizenship competence development for the well-being of young people growing up in today's highly digitized world.

The work of this expert group resulted in the creation and publication of the digital citizenship education handbook. This handbook created a framework that decrypts the end goals of digital citizenship into a language that easily resonates with educators, families, and education policymakers.

The Council of Europe's Competencies for Democratic Culture (CDCs) provides a simplified overview of the competencies that global citizens need to acquire if they are to participate effectively in a culture of democracy. These are not acquired automatically but instead need to be learned and practiced. In this, the role of education is key.

The 20 competencies for democratic culture, cover four key areas: values, attitudes, skills and knowledge, and critical understanding[17] (see Figure 2.2).

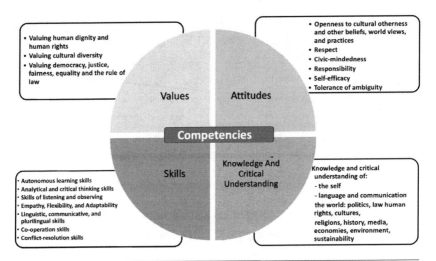

Figure 2.2 The 20 competences for democratic culture.[18]

Digital Domains

To facilitate discussion on the issues and challenges digital citizens encounter in the online world, the Digital Citizenship Education Expert Group divided online activity into three areas: Being online, Well-being online, and Rights online, with 10 supporting digital domains. These domains described below strengthen the overall concept of digital citizenship.

Being online

1. Access and inclusion in digital citizenship emphasize the necessity of not just technical skills but also a strong sense of responsibility and respect grounded in human dignity and rights.

2. Learning and creativity in the digital realm are about adapting education to rapid knowledge turnover and fostering competencies beyond traditional learning.

3. Media and information literacy is defined as the ability to engage meaningfully with media and information channels, critically analyzing and creating content across diverse media.

Well-being online

4. Ethics and empathy in digital citizenship involve understanding and respecting diverse perspectives while demonstrating responsibility toward others, guided by values of human dignity and rights, with empathy acting as a critical component in fostering moral communities and ethical behavior online.

5. Health and well-being emphasize the importance of realizing one's potential and contributing meaningfully to the community, which in the digital context includes managing the stresses and challenges posed by technology to maintain a healthy balance and positive engagement both online and offline.

6. e-Presence and communications are about responsibly managing one's online identity and interactions, understanding the impact of digital footprints, and effectively using communication technologies to enhance personal and communal connections safely and respectfully.

Rights online

7. Active participation involves the competencies needed for citizens to become aware of their role in various environments, enabling them to make informed decisions and contribute positively to democratic cultures both online and offline.

8. Rights and responsibilities encompass understanding and respecting one's online rights, and the accompanying obligations to ensure a fair and safe digital environment for all users.

9. Privacy and security underscores the importance of understanding and safeguarding personal privacy and security online, advocating for responsible behavior that respects the privacy and security rights of all users.

10. Consumer awareness stresses the need for digital citizens to be knowledgeable about their consumer rights and the significance of responsible consumption, including recognizing and responding to the rights and responsibilities associated with digital products and services.

The Five Pillars of Digital Citizenship Development

Five constructs or pillars emerge as being essential in developing effective digital citizenship practices.

These pillars provide a framework for effectively navigating and contributing to the digital world and support the overarching goal of digital citizenship education, which is to equip young citizens with the values, attitudes, skills, and knowledge necessary to participate fully and assume their responsibilities in today's society, both online and offline.

In summary, these five pillars address:

1. Policy: This pillar involves the development of comprehensive policies that guide the responsible use of digital technologies and the Internet, promoting safe and ethical online behavior.
2. Evaluation: This refers to the assessment of digital citizenship initiatives to ensure they are effective and meet the intended goals of educating and empowering digital users.
3. Stakeholders: This encompasses the collaboration between various parties, including educators, parents, policymakers, and platform providers, to support and promote digital citizenship.
4. Strategies: Strategies include the methods and approaches used to implement digital citizenship education, ensuring that it is integrated into various aspects of learning and online engagement.
5. Infrastructure and Resources: This pillar focuses on the availability and accessibility of technical resources, support systems, and educational materials necessary for fostering digital citizenship skills.

While the competencies for democratic culture lay the foundation for digital citizenship, the five pillars uphold the whole structure of digital citizenship development[19] (see Figure 2.3).

The Council of Europe's model for digital competence, often aligned with frameworks like the Digital Competence Framework for Citizens (DigComp), provides a comprehensive structure to assess

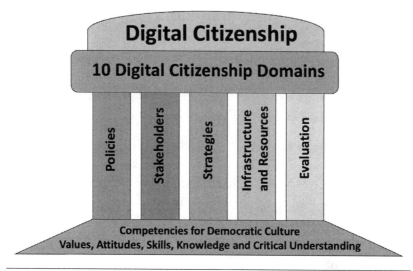

Digital Citizenship

10 Digital Citizenship Domains

Policies

Stakeholders

Strategies

Infrastructure and Resources

Evaluation

Competencies for Democratic Culture
Values, Attitudes, Skills, Knowledge and Critical Understanding

Figure 2.3   The Council of Europe model for digital competence development.[20]

and enhance individuals' digital skills across various dimensions. This model aims to equip citizens with the necessary competencies to participate effectively and responsibly in the digital world.

While the Council of Europe itself focuses on human rights, democracy, and the rule of law in Europe, digital competence models relevant to its values are often designed to foster inclusivity, critical thinking, and safe digital interaction.

The model typically outlines competencies in several key areas:

- Information and Data Literacy: Understanding how to search, locate, evaluate, and use information effectively and critically assess its source and reliability.
- Communication and Collaboration: Knowing how to communicate, interact, collaborate, and share resources digitally, respecting cultural and generational diversity.
- Digital Content Creation: The ability to create, edit, and improve digital content, including text, images, and videos, and understanding copyright and licenses.
- Safety: Awareness and application of safety measures to protect personal data and privacy online. Understanding digital risks, such as cyberbullying and cybercrime, and knowing how to deal with them.

- Problem Solving: Using digital tools to identify, and innovate solutions for personal or community issues, and engaging in lifelong learning through the use of digital technologies.
- Critical Thinking and Ethical Consideration: Evaluating digital content critically and engaging with digital technologies in an ethical, legal, and responsible manner.

This model emphasizes not just the technical skills needed to use digital tools, but also the critical thinking, ethical considerations, and social skills necessary to navigate the digital environment responsibly and effectively. It aims to prepare individuals for a rapidly changing digital world, promoting digital literacy as a foundation for personal development, employment, and citizenship.

The Council of Europe Model: Universal Application

The Council of Europe model for digital competence is designed with a universal application in mind, making it adaptable and relevant across different countries and cultures.

Below are several reasons why The Council of Europe Model may be applied universally and considered by educators as a framework for academic for teaching of digital citizenship to students.

1. Global Digitalization: As digital technologies become increasingly integral to all aspects of society worldwide, the competencies outlined are universally relevant. They address fundamental skills and knowledge needed to navigate the digital world, which is common across global contexts.
2. Inclusivity and Accessibility: The model emphasizes inclusivity and the importance of making digital technologies accessible to all, including marginalized and disadvantaged groups. This universal approach ensures that digital competence initiatives can support broader social inclusion goals.
3. Focus on Fundamental Principles: By centering on core competencies like information literacy, communication, safety, and problem-solving, the model addresses foundational aspects of digital engagement that are applicable regardless of specific technological tools or platforms, which may vary widely between regions.

4. Adaptability to Local Contexts: While the model provides a comprehensive framework, it is flexible enough to be adapted to local educational, cultural, and societal contexts. This adaptability allows countries or regions to tailor the implementation to their specific needs and challenges.

5. Promotion of Critical Thinking and Ethical Engagement: The universal need for critical thinking and ethical engagement in the digital sphere is well recognized. The model's emphasis on these aspects prepares individuals to navigate complex digital environments responsibly, an essential skill in any cultural or national context.

6. Supports Lifelong Learning and Employability: Digital competence is increasingly seen as crucial for lifelong learning and employability, both of which are global challenges. The model's focus on developing skills that enhance personal development and job readiness is universally applicable and beneficial.

By addressing the essential needs of digital engagement and promoting values of inclusivity, responsibility, and adaptability, the Council of Europe's model for digital competence offers a universally applicable framework designed to equip individuals with the skills necessary to thrive in the digital age.

Digital Citizens: Summary

Educating students to become digitally included and competent has to shift away from the consolidated tradition of teaching them how specific software works (thus fomenting operational skills) and move toward educating for competence, thus fomenting skills together with knowledge and attitudes.

This implies the need to be critical and reflective on what we do with technologies, aware of the possibilities and the risks that technologies offer, and ready to move along technological changes to keep up-to-date with the latest developments.[21]

Our Digital Playground: Learning to Be Safe and Kind Online

Teaching children to be good digital citizens and to follow these simple effective principles will help them to develop a cyber-safe approach to actively living, learning, and safely participating in an ever-evolving digitally infused, digitally dependent, society.

Using the acronym **POISE** will aid students in recognizing and remembering the principles of good cyber-safety practices.

A suggested two-step approach to presenting the **POISE** principles would follow…

Step 1, Define and discuss the term poise and provide a recognizable example for students.

Poise means being calm, confident, and controlling your behavior, even when you are in a difficult situation or when people are watching you. It's like being able to stand steady and not wobble, even when you might be feeling a bit nervous inside.

Even though she was nervous during the spelling bee, Maria showed great poise as she took a deep breath and spelled each word correctly.

Step 2, Present the **POISE** principles to students, discussing how these principles will assist them in becoming good cyber-safe digital citizens.

Perceptive means being good at noticing things that are not always clear or easy to see. It's like being a detective and finding clues that help solve a mystery. When you're online, being perceptive can help you notice if something seems strange or if someone is not who they say they are, which can keep you safe on the Internet.

Being **observant** means paying close attention to everything around you. It's like noticing when a friend has a new haircut. If you're observant when you're using the Internet, you can spot when something doesn't seem right, like an email from a friend that doesn't sound like them, which could be a sign that their account may have been hacked. This helps you to avoid clicking on bad links or sharing your passwords.

Impartial means not picking sides and treating everyone the same way. It's like being a referee in a game, making sure that all players are treated fairly, no matter who they are. For example, when you're online, being impartial means, you don't give special treatment to anyone. If someone you don't know asks for personal information, you should be impartial and not share it, just like you wouldn't with anyone else you don't know well, to stay cyber-safe.

Studious means working hard and spending a lot of time learning about something by reading and thinking. It's like when you sit down with your books and concentrate on your homework without getting distracted by other things. For example, being studious about cyber safety means you take the time to learn all the ways to stay safe online, like creating strong passwords, not talking to strangers on the Internet, and always asking a grown-up or trusted adult before downloading something. This way, you're making sure you know how to protect yourself and your computer while you're having fun or studying online.

Engrossed means being so focused on something that you don't notice anything else around you. It's like when you're so into reading a good book that you don't hear someone calling your name. When using technology, if you're engrossed in a game or an app, remember to take breaks and check your privacy settings to make sure you're still safe online.

Lastly and equally important, the principle of being open-minded should be discussed with students when it comes to information and material encountered through the use of technology, be it for academic work or personal enjoyment.

Being open-minded means being willing to listen to and think about new or different ideas. It's like when you try a new food for the first time and decide whether you like it or not after you taste it. In an online world, being open-minded means, you think about the information you see and check if it's true or not before you believe it or share it with others, which helps you be smart and cyber-safe both on- and offline.

Discussing these principles with students will better prepare them to be astute and cyber-safe digital citizens.

Growing Up Digital: A Summary

Discussing the concept of digital citizenship and what it means to be a good digital citizen with students helps to prepare them for the world that awaits them as adults.

A part of guiding young learners toward becoming good digital citizens is to empower them to handle issues that they may encounter when using technology.

To summarize and provide an approach to discussing this topic with your students we recommend the following:

- Explain to students the public nature of the Internet and its risks and benefits. Be sure they know that any digital info they share, such as emails, photos, or videos, can easily be copied and pasted elsewhere and is almost impossible to take back. Remind students that some of this digital communication, like social media posts or photos, could damage their reputation, and friendships, and should not be shared.
- Remind students to be good "digital friends" by respecting the personal information of friends and family, and not sharing anything online about others that could be embarrassing or hurtful.
- Students may face situations like cyberbullying, unwanted contact, or hurtful comments online. Work with them on strategies for when problems arise. These can include talking to a trusted adult right away, refusing to retaliate, calmly talking with the bully, blocking the person, or filing a complaint. Agree on steps to take if the strategy fails. It is better to have these strategies in place ahead of time and be proactive instead of being reactive after an event has occurred.

Teaching digital citizenship to students at a young age helps children hone effective communication skills, allows them to practice good form in social participation, and protects themselves online. It's crucial to learn early about showing respect for others online, recognizing threats, and developing habits that take advantage of the benefits of the Internet while creating resilience to its dangers.[22]

The final word, as the International Society for Technology in Education (ISTE) summarizes so nicely — "Digital citizenship is about more than online safety. It's about creating thoughtful, empathetic digital citizens who can wrestle with important ethical questions at the intersection of technology and humanity"[23] (Figure 2.4).

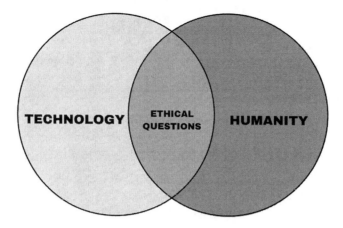

Figure 2.4 Good digital citizens have acquired the necessary skills to address ethical questions at the intersection of technology and humanity.[24]

Basic Cyber Security Concepts for Elementary-Aged Students

Understanding Personal Information

To begin with, broadly, personal information is any information that can be used to distinguish or trace an individual's identity, either alone or when combined with other information that is linked or linkable to a specific individual.

Conveying this concept to elementary-aged children should be done in a simple and relatable manner. For example...

> Kids, today we're going to talk about something called personal information.
> We'll learn about why it's important not to share this information with just anyone and how to protect it, just like we protect our favorite toys or treasures.

What Is Personal Information?

Naturally, personal information will vary in both its breadth and depth based on an individual's age. There are however several items/ identifiers that universally represent personal information. Teaching third- and fourth-grade students to recognize these identifiers and how to keep them safe is a critically important step in developing a cyber-safe approach to engaging and living in a technology-dependent society. A dependency that only grows stronger daily!

What are the most elementary and important personal information identifiers that students should be taught to protect and how can this information be communicated to elementary-aged students?

Consider the following identifiers and an approach to opening discussions with your students...

Imagine personal information like the pieces of a puzzle that can tell someone who you are. This can be your name, where you live, your phone number, or even your favorite color. Just like how we keep certain things private, like a secret diary, we also need to keep our personal information safe. We don't share our personal information with people we don't know well. It's our way of making sure we are safe and happy.

Name   Your name is what people call you. It's like a label on your lunchbox that tells everyone it belongs to you. Your full name is special because it might be shared with family, but be careful who you share that information with, when you are online and especially with someone who you do not know.

Address   Your address is like the name of your house or apartment. It's a special set of words and numbers that tells people exactly where you live. It's like telling your friends which ice cream shop to meet you at, but it's for your home. Again, like your name, be very careful who you share that information with, when you are online and especially with someone who you do not know.

Phone Number   A phone number is like a secret code that connects your phone to you. When someone knows it, they can call you to say "hi" or ask you to play. It's important to only give this code to people you trust, like family and close friends.

Family and Friends   Our family and friends are the people who care about you the most, like your mom, dad, brothers, sisters, aunt, uncle, grandparents, another trusted adult, and your best friends at school. It's okay to talk about who they are but be careful about sharing where they live or their phone numbers, when you are online, with someone you do not know, especially not without asking them first.

Online Games and Apps   Online games are the games you play on the Internet, sometimes with other people from all over the world. They can be really fun, but remember, just because you play with someone online doesn't mean you know them. Be careful not to tell them your personal information.

Apps are like little TV channels on your phone or tablet that let you do certain things like watch videos, play games, or draw pictures. Some apps ask for information about you, but you should always check with a trusted adult before you share anything about yourself.

So, remember, if someone, even if they seem nice, asks for your personal information, always check with a grown-up you trust. They will help you decide if it's okay or not. Remember, personal information is like a secret treasure, and we want to keep it safe!

This list is representative, additional identifiers may be added as appropriate based on student interaction, response, and feedback received by the teacher throughout the class discussion.

Digital Imprint or Digital Footprint?

Before discussing digital footprints, let us take a moment to briefly look at what is called your digital imprint. The terms "digital imprint" and "digital footprint" are often used interchangeably, but they can have slightly different connotations depending on the context.

The term "digital imprint" is less commonly used and can be interpreted in a broader sense. It can refer to the entirety of one's online presence, including both intentional and unintentional elements. A digital imprint could encompass not only the active digital footprints generated by intentional online actions but also the passive elements, such as the information collected by websites and online services without explicit user input.

The concept of a digital imprint might be used more broadly to describe the comprehensive online identity or the collective impact of a person's digital activities.

What's a Digital Footprint?

A digital footprint refers to the trail, traces, or records of a person's online activities. It encompasses the data and information that individuals leave behind while using digital devices, or services, or whenever someone posts information about you on social media platforms (e.g., Facebook, YouTube, TikTok, Snapchat, X, Pinterest, and LinkedIn).

Digital footprints include the websites visited, social media interactions, online searches, posts, comments, and any other online actions that create a traceable record. Digital footprints are often discussed in the context of online privacy and reputation management. They highlight the lasting impact and visibility of one's online actions.

Having a digital footprint in today's digitally infused society is hard to avoid and is pretty normal. The concern is that your digital footprint and all the related personal information associated with that footprint are publicly available.

What types of digital footprints may students leave on the digital landscape? What is the best approach to discussing with students how to act in a cyber-safe manner, to limit their footprints when using technology? These questions and other cyber-safe habits for elementary-age students are discussed in the next section.

Digital Footprints and Elementary-Age Children

Digital footprints and the risks of leaving such footprints in the digital world for everyone to see is an important concept for third- and fourth-grade students to understand. This understanding will help them increase their cyber-safe awareness skills and help mitigate the associated risks when interacting with technology. These skills will be important as they learn to navigate their way through the ever-evolving and evasive world of technology in which they live.

Digital footprints are created in two ways: passively and actively.

A passive footprint is created when your data is collected, usually without you being aware of it. Common examples are search engines storing your search history whenever you're logged in, and web servers logging your computer's IP address when you visit a website.[25]

To discuss the concept of passive footprints with students, consider using the following examples of where students may leave passive digital footprints and how to be cyber-safe in the digital world.

- Online Games and Apps

    Passive Footprint: When playing games or using apps, data about how long a student plays, their favorite activities, or even their location may be collected without them knowing.

Cyber-Safe: Stress the importance of privacy (as discussed previously) and the need to make sure that privacy settings are in place before accessing and playing games. Parents can help too by choosing apps with strong privacy settings and explaining the importance of asking for permission before playing new games. Encourage kids to play offline games as well.

- Smart Toys

Passive Footprint: Interactive toys may record conversations or collect data on a child's preferences to enhance their play experience. It's important to be aware of potential privacy and security concerns. These concerns often stem from the toy's ability to connect to the Internet, collect personal data, or interact with children in ways that could compromise privacy.

Smart Toy Examples:

Interactive Plush Toys (like Furby Connect or Hatchimals)

- Description: These are cuddly stuffed animals that can move, talk, and interact with kids through sensors and voice recognition technology.
- Risks: Since they connect to the Internet to receive updates or download new features, there's a risk that personal conversations could be recorded, or data could be shared without consent. Also, if the toy is hacked, someone could gain unauthorized access to interact with the child.

Educational Tablets (like LeapFrog Epic or Vtech InnoTab)

- Description: These are child-friendly tablets pre-loaded with educational apps, games, and books designed to make learning fun.
- Risks: Tablets often require a connection to the Internet to download new apps or content, which could expose children to inappropriate material if not properly supervised. They may also collect information on usage patterns and personal data.

Remote Control Robots (like Anki Cozmo or Sphero)

- Description: These small robots can be controlled via a smartphone or tablet and are designed to teach coding through play.

- Risks: The apps that control these robots collect data on how they're used. If the robot has a camera or microphone, there's a risk that hackers could use it to see or hear what the robot does, compromising privacy.

Smartwatches for Kids (like VTech Kidizoom Smartwatch)

- Description: These watches not only tell time but also take pictures, play games, and can have controlled communication with parent-approved contacts.
- Risks: They can be connected to the Internet or a smartphone, leading to the potential for location tracking or communication with unapproved contacts if not properly monitored.

Gaming Consoles (like Nintendo Switch or PlayStation with online features)

- Description: These gaming devices allow children to play video games, and many have online capabilities for multiplayer gaming.
- Risks: Online connectivity could expose children to conversations with strangers, online bullying, or sharing of personal information. There's also the risk of making unauthorized purchases through the gaming platform.

When it comes to passive footprints and these smart toys, for example, educators should discuss these potential exposures with students. Additionally, parents and guardians need to research and understand the privacy and security features of smart toys before introducing them to children. This includes understanding data collection practices, the security of Internet or Bluetooth connections, and the company's track record for handling personal information.

Cyber-Safe: A discussion with students could proceed as follows....

Remember, when we play with smart toys that can talk or connect to the Internet, we want to make sure they are safe just like we do when we buckle our seat belts when we ride in the car. So, we always check to make sure our smart toys have their "safety belt" on, which means they have a secret code for example that only you and your family or a trusted adult know.

And just like we keep our name, home address, and phone number private at the playground, and when we are online, we don't tell these things to our smart toys during playtime.

- School Computers and Websites

    Passive Footprint: Discuss with your students how their school computers and educational websites, which they may visit, could track their progress and activities. Some of this collected data may be used for academic or product assessment purposes.

    Cyber-Safe: Discuss with your students the legitimate purpose of software tools that collect information about the sites they visit when they are online and reassure them that their data is used to enhance learning. Emphasize the importance of keeping login information private. Explain that sometimes leaving footprints is alright, unavoidable, and necessary…if done so with the student's knowledge and approval. Once again revisiting the concepts of privacy and personal information.

    You may wish to approach this topic through a discussion with your students that could proceed as follows….

    Hey kids, you know when you play games on the school computers or go to websites to do your homework? Well, these computers and websites are pretty smart. They can remember what games you played, the answers you gave, and even how long you spent on each question. This helps your teachers know how much you're learning and helps make the games and activities better for you.

    But just like superheroes have rules for their powers, we have rules to stay safe online. It's important not to share your passwords with anyone except your parents, a trusted adult, or teachers, and always log out when you're done. Remember, just like we look both ways before crossing the street, we should also be careful and follow the rules when we're exploring the Internet to keep our information safe.

- Photo Sharing

    Passive Footprint: When parents share pictures of their kids on social media, information about the child's interests and activities may be collected. Talk with students about reminding their parents or a trusted adult about the risks of sharing private information.

Cyber-Safe: Educators be mindful of what you may share online about your classroom activities and adjust privacy settings accordingly. Teach kids about the importance of not sharing personal details online. Your actions can be the best cues for acting cyber-safe when using technology. Your students will want to copy your actions!

- Virtual Assistants

    Passive Footprint: Discuss with students to be alert and aware of devices like smart speakers that may record conversations to improve their (the smart speaker's) understanding of user preferences.

    Virtual Assistant Examples

    Amazon Echo (Alexa)
    - Description: Alexa is like a helper who can tell you the weather, play music, or set timers for when you're doing homework.
    - Risks: If Alexa is not set up right, it might listen to private conversations, or someone could use it to buy things without your parents knowing.

    Google Nest
    - Description: Google Nest is a smart speaker that answers questions, helps with homework, and can control lights in your house with your family's permission.
    - Risks: Just like Alexa, if not secured, it could share your family's conversations, or someone might access it from outside to control things in your home.

    Apple HomePod
    - Description: HomePod is a speaker that can play your favorite songs and find answers to your questions by talking to Siri, Apple's voice assistant.
    - Risks: Someone might overhear you if the settings aren't private, and it could let someone else use Siri if it's not locked properly.

    Sonos One (with Alexa or Google Assistant)

- Description: Sonos One is a powerful speaker that can play music from all over the world and uses Alexa or Google Assistant to help you with chores throughout your day.
- Risks: Without the right settings, it might accidentally play music too loud, or someone could change your music without permission.

JBL Link (with Google Assistant)

- Description: JBL Link is a portable speaker that you can carry around and ask questions to Google Assistant, just like you would on a Google Nest.
- Risks: Because it's portable, you have to be extra careful to not lose it and to keep it secure so no one else can use it to find out things about you or your family.

Cyber-Safe: Educators should stress to students that when using these smart speakers, it's really important to make sure they have a password or some kind of lock so that only the student or their family can use them. Also, remind students to turn off the smart devices when not using them, so they don't accidentally hear something they shouldn't. Lastly, emphasize that students should always talk to a grown-up, or a trusted adult if they're not sure how to use a virtual assistant safely!

An active digital footprint is created when you voluntarily share information online. Every time you send an email, publish a blog, sign up for a newsletter, or post something on social media, you're actively contributing to your digital footprint.[26]

To explore the idea of active digital footprints with students. Introduce the concept by providing examples of situations where students may leave active digital traces of their online activities and provide guidance on practicing cyber safety in the digital world.

For example....

- Sharing Photos or Videos

  Active Digital Footprint: Uploading pictures or videos to social media or sharing them with friends online.

  Cyber-safe: Encourage students to check with a trusted adult before sharing any photos or videos. Emphasize the importance of obtaining permission and being aware of what they shared and with whom.

- Posting Personal Information

    Active Digital Footprint: Sharing personal details like their full name, home address, the names of brothers or sisters, or their school on websites or apps.

    Cyber-safe: Teach students to keep personal information private and only share it with their parents or trusted adults. Discuss the concept of online privacy and the importance of not revealing too much about themselves.

- Commenting or Chatting Online

    Active Digital Footprint: Engaging in conversations or leaving comments on websites, games, or social media.

    Cyber-safe: Emphasize the significance of being respectful and kind online. Remind students not to share private information in comments or chats and to report anything that makes them uncomfortable.

- Creating Usernames and Passwords

    Active Digital Footprint: Choosing usernames and passwords for online accounts.

    Cyber-safe: Teach the importance of creating strong and unique passwords. Encourage students to use a mix of letters, numbers, and symbols and to keep their passwords private, only sharing them with their parents or a trusted adult. See the following section "Guardians of Access the Crucial Role of Passwords," for more information to share with your students on creating cyber-safe passwords.

- Online Learning Platforms

    Active Digital Footprint: Participating in online educational activities and submitting assignments.

    Cyber-safe: Students should be taught to follow the teacher's guidelines for online interactions. Reinforce the concept that the Internet is a place for both learning and fun. Remind them that the Internet can also be risky and that it is important to remember and practice their cyber-safety habits when exploring the digital world.

Creating analogies between the physical footprints animals leave and the digital footprints that students create and leave behind in a digital environment can help make explaining the concept of digital footprints more relatable to your students.

Figure 2.5 Digital footprint — posting to social media.[27]

Here are several examples.

- Digital Footprint: Social Media Posts

    Analogy: Displaying your artwork, picture, or painting in your classroom (see Figure 2.5).

    Explanation: Think about when you take a picture or create a drawing in art class and hang it on the wall. Everyone can see it, right? Social media posts are like that but on the Internet. When you post a picture or write something online, it's like hanging your thoughts or pictures for lots of people to see, and they can stay up for everyone to see and read for a very long time. That's part of your digital footprint, and it's how you leave little bits of yourself online. So, always make sure you're sharing happy and kind things that won't make you or others feel upset later.

- Digital Footprint: Online Comments

    Analogy: When you write words on your friend's paper, drawing, or tablet, that is called leaving a comment. Doing this on websites or games is the same, but instead of just your friend, lots and lots of people, all over the world might read what you say (see Figure 2.6).

Figure 2.6   Digital footprint — online comments.[28]

Explanation: Online comments are another piece of your digital footprint. Remember to make your words kind and helpful, not mean and unpleasant just in case someone you know sees or reads them. You want to be proud, not embarrassed, about what you've written and what gets displayed online.

- Digital Footprint: Sharing Personal Information

Analogy: Is like telling your friends and your entire class your secret hiding spot or where you keep your allowance money (see Figure 2.7).

Explanation: Sharing personal information online is like telling everyone your secret and private information. It tells people things that should be private, like your full name, where you live, or your phone number. This is a big part of your digital footprint, and sharing too much can be unsafe. Always check with a grown-up or a trusted adult before you share anything about yourself on the Internet.

- Digital Footprint: Online Gaming/Searching/Social Media

Analogy: When you play online games, search for things on the Internet, or scroll through social media, it's like leaving footprints behind you of where you have been, what you looked at, and what you read (see Figure 2.8).

Figure 2.7 Digital footprint — sharing personal information.[29]

Figure 2.8 Digital footprint — online activities.[30]

Explanation: These "footprints" of information show where you've been and what you like to do. That's your digital footprint in the online world. Just like in real life, you want to leave a good trail, so only go to places that are good and safe for kids, and don't wander off to places that are meant for grown-ups.

Figure 2.9   Digital footprint — email communication.[31]

- Digital Footprint: Email Communication

  Analogy: Your email inbox is like your school locker. It's a private space where you get messages. Just like you have a combination to keep your locker safe, you have a password to keep your email safe so only you can see your messages (see Figure 2.9).

  Explanation: Sending an email is fast because it goes through the Internet. When you write an email, it's like writing a note and pinning it on a friend's private bulletin board where only they can read it. That's part of your digital footprint, too. However, your friend might share your emails with other people. So, when you write an email, make sure that it is polite and sensitive to the recipient and something you wouldn't mind if other people read because, on the Internet, sometimes they do!

Educators, administrators, parents, and trusted adults, whenever possible, should remind students that every time they do something online, it's like putting a sticker with their name on it. Make sure it's a sticker they will be happy for everyone to see!

A Quick Note Regarding Browser Fingerprinting for the Educator

Browser fingerprinting is the act of gathering data from a web browser to create a device's (laptop, mobile phone, etc.) unique fingerprint. This procedure can make remarkable amounts of information about a user's software and hardware environment available, and it can be used to create a special identity termed a browser fingerprint. An analogy with a digital fingerprint is appropriate as the browser fingerprint is often unique to an individual user.

Browser fingerprinting has, without a doubt, become a significant privacy threat to web users. When a user visits a website, browser fingerprinting is used to gather information about system settings and browser configurations, the type and version of the browser, as well as the operating system, IP address, and other current system/user settings.

In the end, a unique identity can be created using this procedure, which can also expose a surprising amount of information about a user's software and hardware settings. The privacy implications are significant because users can then be tracked using these fingerprints.

A third party can identify a person and link their browsing behaviors within and between sessions by gathering browser fingerprints. Most significantly, because the tracking scripts are silent and run in the background, the user is oblivious to the data collection process, which is fully transparent.

Even when the user disables the use of cookies by websites, fingerprints can be used to identify particular users or devices fully or even partially. Many users mistakenly think cookies are browser fingerprints, they are different. The difference is that cookies are only accessible to the website from which they were obtained and cannot be shared from one website to another, whereas browser fingerprinting can track users throughout the Internet.[32]

Discussing Browser Fingerprinting with Elementary-Age Students

The topic of browser fingerprinting is important enough that it should be discussed with third- and fourth-grade students. The discussion, however, will require a simplified, non-technical approach. The following is a suggested approach that educators may use when discussing browser fingerprinting.

Figure 2.10   Browser fingerprinting.[33]

Browser fingerprinting is like when you make a handprint in wet sand or put your fingers in paint and press them on paper. Just like every person's handprint is different, every computer or tablet has its special handprint when connecting to the Internet.

This handprint is made by the computer's information, like what kind of screen it has or what programs it uses, and websites can see this handprint. It helps websites remember your computer, but sometimes we don't want to leave too many handprints everywhere because it can tell people more about us than we want (Figure 2.10).

Remind students to be cyber-safe when online, keep their private information private, and if they ever feel uncomfortable with anything they see online, tell the teacher or a trusted person.

Password Power: The Keys to Keeping Your Online World Safe

Passwords are a first line of defense in protecting the information that is important to you and restricting individuals from accessing data (or information) that they are not authorized to have access to.

Having explained to students the importance of keeping cyber-safe and their personal information private, discussing passwords and the role passwords play in a digital world is an important lesson.

While the subject of passwords and their role in securing our computing devices and the data that reside on them can be technical, explaining the basic concepts to students in the third and fourth grades will require a less technical and straightforward approach.

A recommended approach to beginning the discussion on the critical importance of passwords may proceed as follows...

OK class, today we are going to discuss and explore the need to create strong and secure passwords. Think of a password like a secret code that keeps your clubhouse safe. This code is really important because it makes sure that only you and the people you trust can get inside.

When you're on the computer, playing games, or chatting with friends online, a password does the same thing — it keeps your secrets and personal stuff safe from people you don't know. It's like having an online guard, standing ready to keep your data and information safe and secure (see Figure 2.11).

Figure 2.11 Create and use strong passwords.[34]

Continue explaining to students that it is very important to keep this secret code, secret and safe. If someone finds out their secret code and changes it, they will not be able to unlock (open) their tablet or other electronic devices. Stress that secret codes are to be kept secret, and not tell or give your secret code to anyone unless your teacher, parent, or a trusted person asks you. This way, you can have fun online and be a cyber-safe.

Importance of Strong and Unique Passwords

Now that your students know what a password is, how it works, and that it should be kept secret, discussing how to create a secure password is our next step.

The longer, and the more complex a password is the more difficult it will be for someone to randomly guess it. If the password is complex, the interloper will move on to an easier target. Thus, the objective is to create a strong, complex password that is not time- nor cost-effective for someone to attempt to guess or try to bypass.

Great! That works well for most adults (at least those who follow the rules of creating strong passwords or better yet passphrases) but, what about students in the third and fourth grades?

It is best to begin by explaining that a tricky secret code is better than a simple or easy one and much more difficult for someone to guess. However, for some students remembering their tricky secret code may present a challenge.

The challenge then becomes how to explain to a student the need to create a password that is both secure and at the same time easy to remember. Emphasizing that their secret code and their password should be a little tricky so that only they know it, but not so tricky that they forget it.

A suggested approach is to encourage students to create a password using pictures or drawings. For example, if their password is "BrownBox-9," they can draw a brown box and the number 9 next to the picture. This visual cue helps them remember.

Perhaps some of your students like to play sports creating a password such as "SoccerChamp22." Drawing a soccer ball and writing the number 22 next to the ball provides an important visual clue. This visual clue encourages students to remember their passwords.

For the students who may have a favorite book, helping them to create a password that combines the book title, and a primary color is an approach to creating a strong but easy-to-remember password.

"BlueMoon," from Midnight on the Moon (Magic Tree House #8) The Magic Tree House series by Mary Pope Osborne, for example, could be a good start. Once again, keeping the language simple, using relatable examples, and involving students in the process of creating their special password often leads to successfully memorizing their password, and besides, it is fun!

Reinforce the importance of keeping the password a secret. Explain that just like they don't share their games with strangers, they shouldn't share their passwords either. It's a special secret just for them. Remember that they should only share their password with their teacher, a parent, or a trusted person.

Cybersecurity

Despite the importance of cybersecurity, there is no standard, globally accepted definition for this term.

According to the Cybersecurity and Infrastructure Security Agency (CISA), cyber security is defined as....

> ...the art of protecting networks, devices, and data from unauthorized access or criminal use and the practice of ensuring confidentiality, integrity, and availability of information.[35]

Recognizing the fact that students in third and fourth grade soak up the world around them like a sponge, use this opportunity to explain and whenever possible, demonstrate to your students how you approach keeping your digital world and digital assets safe and secure. They will watch you, listen to you, and mimic you. Knowing this provides a platform for discussing the general concept of cybersecurity with them, whenever the opportunity presents itself.

Once again, using analogies with elementary-age students is an excellent way to approach the broader discussion of cybersecurity. Lead students through a discussion of cybersecurity by beginning the discussion with the following example...

Figure 2.12   Using strong passwords is good cybersecurity.[36]

Imagine you have a castle made of Legos where all your favorite
Lego characters live. You want to keep them safe, so you build
strong walls and maybe even a moat with a drawbridge.

Cybersecurity is like those walls and the moat around your Lego
castle. It helps keep out the bad guys — like dragons or vil-
lains who might want to sneak in and take your Legos or mess
up your castle.

By having good cybersecurity, like using strong passwords, not
talking to strangers online, and keeping your personal information private, you make sure your castle is safe and you can
keep having fun with your Lego adventures (Figure 2.12).

Finally reinforce that as they have learned to keep safe and to look both
ways before crossing the road safely or not talk to strangers, cybersecurity helps us stay safe in the digital world. It's like having a set of rules
and tools to make sure the Internet is a fun and friendly place.

So, when we use computers and tablets to play games or learn new
things, cybersecurity helps make sure everything is protected. It's like
having a moat around your computer, making sure you have a good
time online without any worries!

If technology is to be used in the classroom, there must also be an emphasis on teaching students safe behavior, and how to protect themselves against inherent technology risks.[37]

Summary

The Condition of Education 2023 report published by the National Center for Education Statistics (NCES), disclosed that in 2021, some 97% of 3- to 18-year-olds had home Internet access: 93% had access through a computer, and 28 and 4% relied on a smartphone for home Internet access. The remaining 3% had no Internet access at home.

The overall percentage of 3- to 18-year-olds with home Internet access was higher in 2021 than in 2019 (97% vs. 95%), before the coronavirus pandemic. More specifically, the percentage with home Internet access through a computer was also higher in 2021 than in 2019 (93% vs. 88%). The percentage that relied on a smartphone for home Internet access was lower in 2021 (4%) than in 2016 (5%) and in 2019 (6%).[38]

Crimes against children and youth and the tactics to ensnare them are becoming more sophisticated. The reason... children often use technology-based, Internet-connected devices both in and outside of school.

According to the 2017 School Crime Supplement (SCS) (most recent data available), for the 2016–2017 school year, 20.2% of students reported being bullied in person at school of which 15% (or 3% overall) also reported being bullied electronically — either online or by text.

Students who reported being bullied online or by text reported higher rates of staying home from school (13%) than those students who reported being bullied in person only (3%).[39]

As students use technology to support their learning, schools are faced with a growing need to protect student privacy continuously while allowing the appropriate use of data to personalize learning, advance research, and visualize student progress for families and teachers.[40]

The society and world in which students are living and learning continue to become even more digitally connected and dependent; therefore, the concepts and applications of cybersafety and cybersecurity must be incorporated into the academic curriculum and classroom exercises and through ongoing discussions with young learners about the responsible and safe use of technology.

By nurturing cybersecurity awareness among children, they can become responsible digital citizens, equipped with the expertise needed to protect themselves and contribute to a safer online environment. Early exposure paves the way for the next generation of leaders to manage increasing cyber threats effectively.[41]

LESSON PLANS

Grade 3

TOC Title: G3 Cyber safety
  Lesson Title: Cyber safety
  Grade Level: 3
  Duration: 50 minutes

Objective:

- The students will understand the importance of cyber safety.
- The students will understand basic rules and strategies for staying safe online.
- Students will be able to identify and respond appropriately to potential online risks.

Suggested Materials:

- Whiteboard and markers.
- Examples of safe and unsafe online behavior (included below).
- Printed handouts with basic cyber safety tips and guidelines (included below). Enough copies for groups of two. Cut out and mix up examples so each group has all examples in no particular order.

Procedure

Introduction (10 minutes):

a. Begin with a class discussion on the Internet and its importance in our lives.
b. Ask students what their favorite websites and other activities are online.
c. Explain that even though the Internet is fun and useful, it is important to be safe online, just like in the real world.
d. Introduce the concept of cyber safety — staying safe while using the Internet.

Discussion: Online Safety Rules (15 minutes):

a. Discuss the potential risks of using the Internet, e.g., encountering strangers, seeing inappropriate content, or sharing personal information.

b. Allow students to share any experiences they have had online where they felt uncomfortable or unsafe.

c. Lead a brainstorming session to make a list of rules for staying safe online. Examples might include:
- Never share personal information online, such as your full name, address, or phone number.
- Always ask a trusted adult before downloading anything from the Internet.
- Treat others with respect and never cyberbully.
- Keep passwords private and don't share them with anyone except parents or guardians, and never send them through emails or texts.

d. Write rules on large paper or whiteboard. Leave them posted where students can see them.

Activity (20 minutes):

a. Keep the safety rules posted for students to refer to.
- Divide class into groups of two.
- Pass out sets of online safe and unsafe behavior examples.
- Randomly select one or two examples to do with the class.
- Have groups continue reading and sorting the examples.
- Encourage discussion between students about why a choice is safe or unsafe.
- If time allows, ask students to come up with alternative safe behaviors for the unsafe examples.

Conclusion (5 minutes):

a. Review the key points of cybersafety.
b. Distribute printed handouts summarizing cybersafety tips and guidelines for students to take home (included below).
c. Encourage students to share what they learned with their families and to ask for help if they ever feel unsafe online.

Assessment:

a. Informally assess students' understanding throughout the lesson by observing their participation in discussions and activities.

b. At the end of the lesson, ask students to share one thing they learned about staying safe online.

Homework (optional):

a. Encourage students to share what they learned about cyber safety with their families.

Safe and Unsafe Online Behavior Examples:

Safe Online Behavior:

1. Using a Kid-Friendly Search Engine: Searching for information about dinosaurs for a school project using a kid-friendly search engine like Kiddle or KidzSearch.
2. Playing Online Games with Friends: Playing an online game with friends from school, using a username that doesn't reveal personal information.
3. Asking Permission Before Sharing: Asking a parent or guardian before posting a photo or video online, even if it's just to share with family and friends.
4. Reporting Inappropriate Content: Reporting a message or comment that is mean or makes them feel uncomfortable to a trusted adult or the website's moderator.
5. Being Kind Online: Leaving a positive comment on a friend's artwork or sharing a funny joke with classmates in a safe online chat room.

Unsafe Online Behavior:

1. Talking to Strangers: Chatting with someone they don't know online without checking with a parent or guardian first, even if they seem friendly.
2. Sharing Personal Information: Giving out their full name, address, or phone number in an online game or chat room without realizing it could be dangerous.
3. Clicking on Unknown Links: Clicking on a pop-up ad promising free games or prizes without realizing it could lead to downloading a virus or scam.

4. Bullying or Being Mean: Sending a mean message to another student or making fun of someone online, even if it's just as a joke.

5. Downloading Without Permission: Downloading a new app or game without asking a parent or guardian first, which could lead to accidentally spending money or downloading inappropriate content.

TOC Title: G3 Digital Citizenship
Lesson Title: Digital Citizenship
Grade Level: 3
Duration: 45 minutes

Objective:

- The students will understand the concept of digital citizenship.
- The students will be able to demonstrate responsible and safe behavior online.

Suggested Materials:

- Whiteboard and markers
- Printed handouts with examples of online scenarios (provided below)

Procedure

Introduction (10 minutes):

a. Whole group discussion: ask students what the term "digital citizenship" means. Write down their responses on the whiteboard.
b. Explain that digital citizenship means the responsible and safe use of technology, including the Internet.
c. Talk about why it is important to be good digital citizens. You can mention concepts like cyberbullying, privacy, and respecting other people online.

Activity (25 minutes):

a. As a class, go over some different scenarios related to digital citizenship, such as:
  - If someone you don't know tries to add you as a friend online, what should you do?
  - How should you respond if someone is mean to you in a chat or comments section?
  - What information should you keep private when you are online?

b. Divide the students into groups of 3–4 students each and give each group a printed handout with one of the scenarios. Do not give them the answer. Have groups discuss what they would do in that situation and come up with a solution.

c. Reconvene as a class and have each group share their thoughts and solutions. In a constructive, positive manner, compare group solutions with given answers. Are the answers similar? If not, are they both good responses? Is one response better? Why?

Conclusion (10 minutes):

a. Summarize the key points of good digital citizenship, e.g., being kind, respectful, and responsible online.

b. Ask students to reflect and share what they have learned and share one thing they will do to be a better digital citizen.

Assessment:

a. Observe students' participation in discussions and group activities. Assess their understanding of digital citizenship based on their contributions and reflections during the lesson.

Homework (optional):

a. Encourage students to share what they learned about digital citizenship with their families.

Digital Citizenship Scenarios and Answers:

Scenario 1: Chat Room Conversation:

While in an online chatroom, someone you don't know starts asking you personal questions like where you live and what school you go to. How should you respond?

Answer: Don't share personal information with strangers online. Consider leaving the chatroom if someone makes you uncomfortable. Talk to a trusted adult about the situation.

Scenario 2: Online Purchases:

You find a website selling a toy you really want, but it looks different from other online stores you've used before. What should you check before making a purchase?

Answer: Check the website's reviews and ratings, as well as its security features (like https:// - not just http:// in the URL) before making a purchase. Only buy from trusted and secure websites.

Scenario 3: Fake News:

You see an article shared online claiming that eating a certain food will make you super smart. However, you've never heard of the website before. What should you do before sharing the article?

Answer: Verify the credibility of the website by checking if it's a reputable source before sharing the article. Look for other sources to confirm the information.

Scenario 4: Password Sharing:

Your friend asks for your password to a website so they can log in and play a game. What should you tell your friend?

Answer: Explain to your friend that sharing passwords is not safe and could put both of your accounts at risk. Encourage them to create their own account.

Scenario 5: Cyberbullying:

You see someone being mean to another student in a group chat. They're saying hurtful things and making fun of the other students. What should you do?

Answer: Speak up against cyberbullying by telling the person to stop or reporting the behavior to a teacher or parent. Be supportive of the person being targeted.

Scenario 6: Privacy Settings:

You realize that your profile on a social media site is set to public, meaning anyone can see your posts and information. What should you do to protect your privacy?

Answer: Adjust your privacy settings to make your profile private so only friends can see your posts and information. Review and update privacy settings regularly.

Scenario 7: Clickbait:

You come across a link that promises a free game if you click on it, but it looks suspicious. What should you do?

Answer: Avoid clicking on suspicious links, especially if they seem too good to be true. It's better to be safe than sorry.

Scenario 8: Sharing Photos:

You take a funny picture of your friend and want to post it online. However, your friend doesn't want you to share it. What should you do?

Answer: Respect your friend's wishes and don't share the photo if they don't want you to. Always ask for permission before sharing someone else's picture.

Modify or expand upon these scenarios based on the specific needs of your students.

Grade 4

TOC Title: G4 Cyber safety
  Lesson Title: Cyber safety
  Grade Level: 4
  Duration: 50 minutes

Objective:

- The students will be able to identify potential online risks.
- The students will be able to employ strategies for staying safe online.

Suggested Materials:

- Whiteboard and markers.
- List of cyber safety scenarios for discussion groups (Provided).

Procedure

Introduction (5 minutes):

a. Whole group discussion about students' experience with the Internet. What websites do students visit? What do they do online?
b. Explain that this lesson is about cyber safety which means staying safe while using the Internet.

Discussion of Online Safety Rules (10 minutes):

a. Lead a class discussion on the potential risks of being online. Ask students if they can share any experiences they have had or are aware of when someone had a bad or unsafe experience online.
b. Write their responses on the whiteboard.

Cyber Safety Rules (10 minutes):

a. Introduce a set of cyber safety rules that students should follow when online. These rules might include:
   - Never share personal information online, e.g., full name, address, phone number, or school.
   - Always ask a parent or guardian before downloading anything or making any online purchases.
   - Always treat others with kindness and respect.

- Report any messages or content that seem inappropriate or make you uncomfortable.

b. Add any other rules the class comes up with that are appropriate.

c. Discuss each rule and explain why it is important to follow them.

d. Post the rules where students can easily refer to them.

Activity (20 minutes):

a. Divide students into small groups (3–4 students per group) and give each group the scenarios list provided. Leave out the discussion points portion.

b. Each group reads and discusses the scenarios and comes up with a safe response.

c. Groups share their responses with the class.

d. Point out similarities or differences in the groups' responses.

e. Use the discussion points included in the scenarios list to guide groups if needed.

Conclusion (5 minutes):

a. Review key points of the lesson.

b. Remind students to always follow the cyber safety rules.

Assessment:

a. Monitor students' participation in the discussion and their ability to apply cyber safety rules to the interactive activity.

Homework (optional):

a. Encourage students to share what they learned about cyber safety with their families.

Scenarios for Group Discussions:

Online Friend Requests:

- Scenario: You receive a friend request on a social media platform from someone you do not know. They claim to be a friend of a friend. What should you do?
- Discussion Points: Teach children not to accept friend requests from strangers and to talk to a trusted adult if they receive such requests.

Sharing Personal Information:

- Scenario: You are playing a game online and someone asks you for his address so they can send you a gift. What should you do?
- Discussion Points: Emphasize the importance of never sharing personal information online, including full name, address, phone number, or school name.

Dealing with Cyberbullying:

- Scenario: You see mean comments about your friend on a group chat. What should you do?
- Discussion Points: Teach children to stand up to cyberbullying by either reporting it to a trusted adult or using the platform's reporting tools. Encourage empathy and kindness online.

Password Protection:

- Scenario: Your friend wants to know your password so they can play a game with you. What should you do?
- Discussion Points: Discuss the importance of keeping passwords private and not sharing them with anyone, even friends. Teach the importance of creating strong passwords.

Clicking on Unknown Links:

- Scenario: You receive an email with a link promising a free game download. What should you do?
- Discussion Points: Teach children to be cautious about clicking on links from unknown sources, as they could contain a

virus or may lead to an unsafe website. Encourage students to ask a trusted adult before clicking on any unfamiliar links.

Inappropriate Content:

- Scenario: You accidentally come across a website with inappropriate content while searching for information for a school project. What should you do?
- Discussion Points: Teach children to exit out of the website immediately and talk to a trusted adult about what they saw. Encourage them to be cautious about the websites they visit and to use safe search settings when browsing the Internet.

Online Gaming Safety:

- Scenario: You are playing a game online and someone you do not know starts asking you personal questions. What should you do?
- Discussion Points: Teach children to immediately stop communicating with strangers online. Block and report them if needed. Encourage them to only play games with people they know in real life or with parental supervision.

Online Purchases:

- Scenario: You find a website selling your favorite toys at very cheap prices. Purchases require a credit card. What should you do?
- Discussion Points: Teach children to never make online purchases without permission from a trusted adult. Emphasize the importance of verifying the legitimacy of websites before entering any payment information.

Location Sharing:

- Scenario: You are using a social media app that asks for your location to tag your posts. Should you share your location?
- Discussion Points: Discuss the risks of sharing location information online and teach children to be cautious about revealing their whereabouts to strangers. Encourage them to disable location services on social media apps.

Downloading Apps:

- Scenario: You want to download a new game app, but it is asking for permission to access your camera and contacts. Should you download the app?
- Discussion Points: Teach children to review app permissions carefully before downloading anything. Emphasize the importance of only granting permissions that are necessary for the app's function.

TOC Title: G4 Digital Citizenship
Lesson Title: Digital Citizenship
Grade Level: 4
Duration: 45 minutes

Objective:

- The students will understand the concept of ethical digital citizenship.
- The students will be able to demonstrate responsible online behavior.

Suggested Materials:

- Printed copies of the Digital Citizenship Pledge (one per student — sample included below).
- List of role-playing scenarios (provided below).
- Paper and colored pencils/markers.
- Chart paper and markers.

Procedure

Introduction (5 minutes):

a. Whole group: ask students if they know what digital citizenship is.
b. If students do not already know the term, explain that digital citizenship is the responsible and ethical use of technology, including things like the Internet, and social media.
c. If students are unfamiliar with the term ethical, give a definition in age-appropriate terms. Examples might include doing the right thing, making good choices, or being kind, honest, and fair to everyone even when no one is watching.

Brainstorm Activity (10 minutes):

a. Divide the class into groups of 3–4 students, adjusting as needed for class size.
b. Provide each group with chart paper and markers.
c. Have groups brainstorm examples of good and bad online behavior and write them down.

d. After 5 minutes, ask each group to share their examples and briefly discuss why each behavior is either responsible or irresponsible.

Digital Citizenship Pledge (10 minutes):

a. Distribute printed copies of the Digital Citizenship Pledge to each student (sample pledge provided below).
b. Read the pledge aloud together as a class.
c. Encourage students to reflect on the pledge and discuss why each statement is important.
d. Have students sign their names on the pledge as a commitment to being responsible digital citizens.

Activity (15 minutes) Role Playing Scenarios:

a. Have students pair up with a partner or assign partners if necessary.
b. Provide each pair of students with a scenario from the list provided. Do not give students the "Ethical Response" portion.
c. Ask students to role-play the scenario, demonstrating how they would respond responsibly.
d. Afterward, facilitate a discussion about the strategies used and alternative ways to handle each situation. If needed, use the "Ethical Response" suggestions provided to help any groups who are struggling.

Conclusion (5 minutes):

a. Review the key points discussed during the lesson.
b. Remind students of their Digital Citizenship Pledge.
c. Encourage students to apply what they have learned in their everyday digital interactions.

Assessment:

a. Review students' completed Digital Citizenship Pledge to ensure understanding and commitment.
b. Evaluate students' role-play performances based on their demonstration of responsible online behavior and effective communication skills.

Homework (Optional):

a. Encourage students to share what they learned about cyber safety with their families.

Digital Citizenship Pledge:

As a responsible digital citizen, I pledge to:

1. Always be kind and respectful to others online.
2. Think before I post or share anything online. I will ask myself how this might affect others.
3. Protect my personal information and privacy by being cautious about what I share online.
4. Stand up against cyberbullying and report any harmful or inappropriate behavior I encounter.
5. Use technology safely and appropriately for my age.
6. Always respect the rights and feelings of others.
7. Be honest and trustworthy in all of my online interactions.
8. Seek help from a trusted adult if I ever feel uncomfortable or unsafe online.

I understand that being a responsible digital citizen means making good choices and treating others with kindness and respect online. By signing this pledge, I commit to being a positive and ethical member of the digital community.

[Student's Signature] _____

Date: _____

Role Playing Scenarios and Suggested Ethical Responses:

1. **Mean Comment:** One student posts a mean comment on another student's social media profile picture making fun of their appearance. The other student sees the comment and feels hurt and embarrassed. How would you respond to the mean comment?

   Ethical Response: The student should not engage with the mean comment. Instead, they should consider blocking and reporting the person who made the comment. They can also talk to a trusted adult about how the comment made them feel.

2. **Inappropriate Content:** While searching for information online for a school project, a student comes across a website with inappropriate content, e.g., violent images, explicit language, etc. What should the student do?

   Ethical Response: The student should immediately close the website and inform a trusted adult about what they saw. They should avoid sharing inappropriate content with others and be cautious when browsing online in the future.

3. **Cyberbullying:** A classmate's friend starts spreading rumors about them online, making hurtful comments and posting embarrassing photos. How should your classmate handle this situation?

   Ethical Response: They should save evidence of cyberbullying (screenshots, messages) and report it to a trusted adult. They should not retaliate or respond to the cyberbully directly.

4. **Sharing Personal Information:** A student receives a message from someone they don't know online, asking for their address and phone number. How should the student respond to this request?

   Ethical Response: Never share personal information with strangers online. They should politely decline the request and inform a trusted adult about the message they received.

5. **Online Gaming Etiquette:** While playing a game online, a player on your team starts insulting and mocking others including you. How should you respond to this behavior while still enjoying the game?

Ethical Response: Mute or block the player who is being disrespectful and report their behavior to the game's moderators. Focus on playing the game and not engage with the player's negative behavior.

6. **Responding to a Chain Message:** A student receives a chain message from a friend claiming that something bad will happen if they don't forward it to ten other people. How should the student respond to this message?

     Ethical Response: The student should ignore the chain message and not forward it to others. They can explain to their friend that chain messages are not reliable and can spread misinformation or cause unnecessary worry.

7. **Online Challenges:** A student sees a viral online challenge that involves dangerous or risky behavior. They feel pressured to do it to fit in with their peers. What should the student do?

     Ethical Response: The student should recognize the dangers of participating in such challenges and refuse to take part. They can educate others about the risks involved and encourage them to make safer choices.

8. **Sharing Passwords:** Your friend asks for your password to access a gaming or social media account. They promise they will not misuse it, but insist that they need it. What should the student do?

     Ethical Response: Never share your passwords with anyone, even a friend. Explain to your friend the importance of keeping passwords private and suggest alternative ways to play or interact online.

9. **Online Friend Request:** A student receives a friend request from someone they don't know. The person claims they have mutual friends, but it seems suspicious. What should the student do?

     Ethical Response: The student should decline the friend request and block the person. They should only accept friend requests from people they know and trust in real life.

10. **Online Purchases:** A student finds a website selling items at very low prices for a limited time. They are excited about the deals and want to purchase items, but would need to use their mom's credit card without her permission. What should the student do?

   Ethical Response: The student should not make any purchases without their parent's permission. They should discuss the website with their parents and make sure it is legitimate before making any transactions online.

Notes

1 Nevens, T. M., winter 2001, "Fast Lines at Digital High," *The McKinsey Quarterly*, pp. 167–177. Gale Academic OneFile, link.gale.com/apps/doc/A72524631/AONE?u=anon~3971fa9e&sid=googleScholar&xid=3 be7d694, retrieved February 1, 2024.

2 The overview of the pervasiveness of technology was developed based on a review and summarization of multiple documents and supported in part, through references from the following sources: (a) Haleem, A., Javaid, A., Qadri, M., & Suman, R., May 23, 2022, "Understanding the role of digital technologies in education: A review," *Sustainable Operations and Computers*, pp. 275–285, Published by Elsevier B.V. on behalf of KeAi Communications Co., Ltd. This is an open-access article under the CC BY license (http://creativecommons.org/licenses/by/4.0/), (b) Lucke, U., & Rensing, C., October 2014, "A Survey on Pervasive Education," *Pervasive and Mobile Computing*, 14, 3–16, ISSN 1574-1192, (c) Cladis, A., September 2020, "A Shifting Paradigm: An Evaluation of the Pervasive Effects of Digital Technologies on Language Expression, Creativity, Critical Thinking, Political Discourse, and Interactive Processes of Human Communications," *E-Learning and Digital Media*, 17(5), 341–364, https://eric.ed.gov/?id=EJ1260987, and (d) Shubina, I., & Kulakli, A., 2019. "Pervasive Learning and Technology Usage for Creativity Development in Education," *International Journal of Emerging Technologies in Learning (iJET)*, 14, 95. https://doi.org/10.3991/ijet.v14i01.9067, retrieved January 10, 2024.

3 (n.a.), March 29, 2023, "Children and parents: media use and attitudes report 2023," Ofcom, www.ofcom.org.uk/research-and-data/media-literacy-research/childrens/children-and-parents-media-use-and-attitudes-report-2023, retrieved February 15, 2024.

4 (n.a.), February 22, 2021, "Middle Childhood (6–8 years of age): Important milestones for children aged 6–8:" Centers for Disease Control and Prevention, www.cdc.gov/ncbddd/childdevelopment/positiveparenting/middle.html, retrieved February 15, 2024.

5 (n.a.), September 23, 2021, "Middle Childhood (9–11 years of age): Important milestones for children aged 9–11," Centers for Disease Control and Prevention, www.cdc.gov/ncbddd/childdevelopment/positiveparenting/middle2.html, retrieved February 15, 2024.

6 (n.a.), March 29, 2023, "Children and parents: media use and attitudes report 2023," Ofcom, www.ofcom.org.uk/research-and-data/media-literacy-research/childrens/children-and-parents-media-use-and-attitudes-report-2023, retrieved February 15, 2024.

7 (n.a.), March 29, 2023, "Children and parents: media use and attitudes report 2023," Ofcom, www.ofcom.org.uk/research-and-data/media-literacy-research/childrens/children-and-parents-media-use-and-attitudes-report-2023, retrieved February 15, 2024.

8 (n. a.), 2022, "Technology in Teaching and Learning," EdWeek Research Center, https://epe.brightspotcdn.com/8d/b6/49769ee54be 9af7ed5287b6b2a0a/technology-in-teaching-and-learning-research-spotlight-4.13.22_Sponsored.pdf, retrieved February 16, 2024.

9 (n. a.), January 25, 2023, "2023 Trends in K–12 Education," Hanover Research, www.hanoverresearch.com/reports-and-briefs/2023-trends-i n-k-12-education/?org=k-12-education, retrieved February 12, 2024.

10 Thomas, S., et al., October 2016, "Policy Brief on Early Learning and Use of Technology," U.S. Department of Education, Office of Educational Technology, Washington, DC, http://tech.ed.gov/ear-lylearning, retrieved February 13, 2024.

11 Karppinen, I., Nurse, J. R. C., & Varughese, J., 2023, "Oh Behave! The Annual Cybersecurity Attitudes and Behaviors Report 2023," The National Cybersecurity Alliance and CybSafe, www.cybsafe.com/ whitepapers/cybersecurity-attitudes-and-behaviors-report, retrieved January 28, 2024.

12 Snyder, S., 2016, "Teachers' Perceptions of Digital Citizenship Development in Middle School Students Using Social Media and Global Collaborative Projects," Walden Dissertations and Doctoral Studies, https://scholarworks.waldenu.edu/dissertations/2504, retrieved January 19, 2024.

13 (n. a.), January 2017, "Reimagining the Role of Technology in Education: 2017 National Education Technology Plan Update," U.S. Department of Education, Office of Educational Technology, Washington, DC, Department of Education, http://tech.ed.gov, retrieved February 2, 2024.

14 Richardson, J., & Milovidov, E., 2019, "Digital citizenship education handbook: Being online, well-being online, and rights online," © Council of Europe, reproduced with permission, https://rm.coe. int/16809382f9, retrieved February 26, 2024.

15 Ferrari, A., Editors: Punie, Y., & Brečko, B., 2013, *DIGCOMP: A Framework for Developing and Understanding Digital Competence in Europe*, European Commission, Joint Research Centre, Institute for Prospective Technological Studies, ISBN 978-92-79-31465-0, retrieved February 22, 2024.

16 (n. a.), January 2017, "Reimagining the Role of Technology in Education: 2017 National Education Technology Plan Update," U.S. Department of Education, Office of Educational Technology, Washington, DC, Department of Education, http://tech.ed.gov, retrieved February 3, 2024.

17 Richardson, J., & Milovidov, E., 2019, "Digital citizenship education handbook: Being online, well-being online, and rights online," © Council of Europe, reproduced with permission, https://rm.coe. int/16809382f9, February 26, 2024.

18 Richardson, J., & Milovidov, E., 2019, "Digital citizenship education handbook: Being online, well-being online, and rights online," © Council of Europe, reproduced with permission, https://rm.coe.int/16809382f9, February 26, 2024.

19 Richardson, J., & Milovidov, E., 2019, "Digital citizenship education handbook: Being online, well-being online, and rights online," © Council of Europe, reproduced with permission, https://rm.coe.int/16809382f9, February 26, 2024.

20 Richardson, J., & Milovidov, E., 2019, "Digital citizenship education handbook: Being online, well-being online, and rights online," © Council of Europe, reproduced with permission, https://rm.coe.int/16809382f9, February 26, 2024.

21 Ferrari, A., Neža Brečko, B., Punie, Y., May 2014, "DIGCOMP: a Framework for Developing and Understanding Digital Competence in Europe," eLearning Papers, ISSN: 1887-1542, www.openeducationeuropa.eu/en/elearning_papers, retrieved February 22, 2024.

22 (n. a.), 2023, "Digital Citizenship for Students," Rachel's Challenge, https://rachelschallenge.org/get-info/digital-citizenship, retrieved January 29, 2024.

23 (n. a.), (n.d.), ISTE Standards, International Society for Technology in Education (ISTE), https://iste.org/standards, retrieved February 15, 2024.

24 Figure drawn by the authors based upon, the International Society for Technology in Education (ISTE) standards.

25 (n. a.), 2016, "My Digital Footprint: A Guide to Digital Footprint Discovery and Management," Centre for the Protection of National Infrastructure, National Protective Security Authority, www.npsa.gov.uk/resources/my-digital-footprint-brief-guide, retrieved February 2, 2024.

26 (n. a.), 2016, "My Digital Footprint: A Guide to Digital Footprint Discovery and Management," Centre for the Protection of National Infrastructure, National Protective Security Authority, www.npsa.gov.uk/resources/my-digital-footprint-brief-guide, retrieved February 2, 2024.

27 Shutterstock image. Stock Vector ID 2232253007. Used under license from CRC Press.

28 Shutterstock image. Stock Vector ID 2240084589. Used under license from CRC Press.

29 Shutterstock image. Stock Vector ID 2416841483. Used under license from CRC Press.

30 Shutterstock image. Stock Vector ID 2398295365. Used under license from CRC Press.

31 Shutterstock image. Stock Vector ID 2118346526. Used under license from CRC Press.

32 Pau, K. N., Lee, V. W. Q., Ooi, S. Y., & Pang, Y. H., March 13, 2023, "The Development of a Data Collection and Browser Fingerprinting System," *Sensors (Basel, Switzerland)*, 23(6), 3087. https://doi.org/10.3390/s23063087, www.ncbi.nlm.nih.gov/pmc/articles/PMC10057587/, retrieved March 1, 2024.

33 Shutterstock image. Stock Photo ID 549761713. Used under license from CRC Press.

34 Shutterstock image. Stock Vector ID 1651382374. Used under license from CRC Press.

35 (n. a.), February 1, 2021, "What is Cybersecurity?" Cybersecurity and Infrastructure Security Agency (CISA), www.cisa.gov/news-events/news/what-cybersecurity, retrieved March 1, 2024.

36 Shutterstock image. Stock Vector ID 93717034. Used under license from CRC Press.

37 (n. a.), (n.d.), "The Risks of Technology as an Aid in K-12 Education," Safe Search Kids, www.safesearchkids.com/the-risks-of-technology-as-an-aid-in-k-12-education, retrieved March 1, 2024.

38 Irwin, V., Wang, K., Tezil, T., Zhang, J., Filbey, A., Jung, J., Bullock Mann, F., Dilig, R., & Parker, S. (2023). Report on the Condition of Education 2023 (NCES 2023–144rev). U.S. Department of Education. Washington, DC: National Center for Education Statistics, https://nces.ed.gov/pubsearch/pubsinfo.asp?pubid=2023144rev, retrieved March 1, 2024 and National Center for Education Statistics. (2023). Children's Internet Access at Home. Condition of Education. U.S. Department of Education, Institute of Education Sciences, https://nces.ed.gov/programs/coe/indicator/cch, Retrieved March 1, 2024.

39 Yanez, C., Seldin, M., & Rebecca, M., November 2019, This National Center for Education Statistics (NCES) Data Point, Synergy Enterprises, https://nces.ed.gov/pubs2020/2020042.pdf, retrieved March 2, 2024.

40 (n. a.), (n.d.), National Education Technology Plan (NETP), U.S. Department of Education, https://tech.ed.gov/netp/introduction, retrieved March 2, 2024.

41 (n. a.), (n.d.), "Teaching Kids About Cybersecurity: Engaging Methods for Young Minds," Safe Search Kids, https://www.safesearchkids.com/teaching-kids-about-cybersecurity-engaging-methods-for-young-minds/#:~:text=By%20nurturing%20cybersecurity%20awareness%20among,manage%20increasing%20cyber%20threats%20effectively, retrieved March 1, 2024.

3

STRANGERS

Safe/Unsafe People and the Tricky Person

This chapter focuses on conveying the importance of knowing who a safe person is and why not all people a student may know, or nice people, in general, are "safe people."

This chapter focuses on educating children about the distinction between safe and unsafe individuals, including the concept of "tricky people." It emphasizes the importance of children understanding who can be trusted and the limitations of the "stranger-danger" approach. The chapter introduces strategies for discussing safety with children, including recognizing safe adults, understanding body autonomy, and establishing family safety protocols. Key topics include identifying safe and unsafe people, online and offline, the concept of tricky people, setting boundaries, and proactive communication strategies for children to express discomfort or seek help.

Introduction

As children grow and become more independent, personal safety becomes a more important concern. Interpersonal interactions become more frequent and less supervised. We move from discussing character education and nurturing digital citizens to another equally important area, yet one that may be difficult to explain to younger students who are gaining increasingly more independence.

Who Is a Tricky Person?[1]

Previously, danger was an unknown bad guy, who presented *him*self in person and was named stranger danger.

DOI: 10.1201/9781003466338-3

The threat has moved from strictly an in-person encounter to include an online encounter. There is often a hybrid version where the online meeting is nurtured and is subsequently used to gain trust to facilitate an in-person meeting. This hybrid created the term tricky person.

At one time, it was considered that danger came in the form of a stranger, but we now know that danger may also be encountered in the form of a person, or someone known to the child, and often known to the family. Someone may be a stranger, but not a dangerous person, and is, a helpful and needed person, like a police officer or a firefighter. At this age, children are starting to distinguish between the helpful stranger and how to handle an emergency.

The idea of a tricky person is to convey to the child that there is deception in trust. In tandem with the deception and trust concept, the term safe or unsafe person is also used. Using both tricky and unsafe people speaks to actions and less about the relationship involved (see Figure 3.1).

According to the Centers for Disease Control and Prevention, someone known and trusted by the child or child's family members perpetrates 91% of child sexual abuse.[2] This leaves only 8% perpetrated by a person unknown to the child.

Who Is and Is Not a Tricky or Unsafe Person?

How to talk with children at the earliest ages about a tricky person and deception, and who is a safe or unsafe person?

(a)   (b)

Figure 3.1 Safe person vs tricky person.[3]

The website Lifehacker's, article on teaching children about safe people with TikTok creator and parenting expert, Jessica Martini, suggests sharing the following parameters with kids to identify safe adults in their lives, rather than creating a traditional "list" by names or titles. These parameters are good to practice when talking about safety with children.[4]

- We can begin by telling our kids: "A safe adult is someone who makes you feel happy and safe," Martini explains. "When you go near them, you don't feel nervous, scared, or have an icky feeling in your tummy. They make you feel loved and comfortable."

- Secondly, a safe adult (or adolescent) "will never, ever ask you to keep a secret," and if they do, instruct your child to share the secret with you right away. She also advises not to teach kids about "good vs. bad secrets" (ones that make them feel happy vs. bad) because sometimes abusers will use "good secrets to get their foot in the door." (e.g. "Here's a cookie, don't tell your mom" or "I know you broke that, but I won't tell anyone, it's our little secret." Getting kids in the habit of keeping "good secrets" makes it easier for them to keep bad secrets in the future. Instead, Martini notes, we can teach them the difference between secrets and "happy surprises" — things that everyone will find out soon, and everyone will feel happy about, such as a surprise party.

- A safe adult will *always believe you* when you tell them something important. Martini points out that sometimes our kids will "tell us without telling us" when they don't have the words or emotional maturity to convey it verbally. They may act out the traumatic situation in their play, experience headaches, stomach aches (sic), or other signs of physical illness (especially when they're going to see a certain person), become withdrawn, or experience a loss of appetite. They may also act distressed when going somewhere they previously enjoyed going — exhibiting behavior such as screaming, crying, clinging, refusing to move, or saying "They're the worst, I never want to go there."

Kids First, Inc., started in 1991 and serving up to 250 children per year, provides the following list of subjects to discuss with children about body safety and the safe/unsafe person concept.

- Talk about "safe" and "unsafe" touching rather than "good" or "bad" touching. This removes guilt from the child and keeps them from having to make a moral distinction about what is and is not appropriate.
- Use age-appropriate wording. You can discuss body safety without discussing sexuality. Teach young children that no one should touch them in any area that their bathing suit covers, and that they should never touch anyone else in these areas or see pictures or movies that show those areas.
- Teach the difference between healthy and unhealthy secrets. An example is that a surprise party is an okay secret to keep because it will make people happy and will be told at the right time. Secret touching is not okay or keeping any permanent secrets from parents or caregivers.
- Have your child identify five people that they could/would talk to if someone was touching them in an unsafe way. Children are often afraid to tell their parents out of fear of punishment (or because of a threat made by a perpetrator), so your child needs to know they can seek out other trusted adults to confide in. Instruct your child that they should keep telling until someone helps them.
- Teach children proper names for body parts so that if they disclose inappropriate touching, it will be clear what they are talking about.
- Revisit this safety talk often. Children learn through repetition. How many times do you remind children to look both ways before crossing the street? (Figures 3.2–3.5).[5]

Planet Puberty, a digital resource suite by Family Planning NSW, aims to provide parents and carers of children with intellectual disability and/or autism spectrum disorder with the latest information, strategies, and resources for supporting their child through puberty. Planet Puberty's focus is on children with intellectual disability and

> ## Child Safety Disclosure What to Say
>
> - Thank you for telling me.
> - I believe you.
> - I will support you.
> - Everyone has the right to feel safe.
> - There is nothing so bad that you can't tell someone.

Figure 3.2   What to say.[6]

> ## Child Safety Disclosure What to Say
>
> - I am sorry that happened.
> - Thank you for trusting me.
> - It is not your fault.
> - You are brave telling me.
> - You did the right thing telling me.

Figure 3.3   What to say (continued[7]).

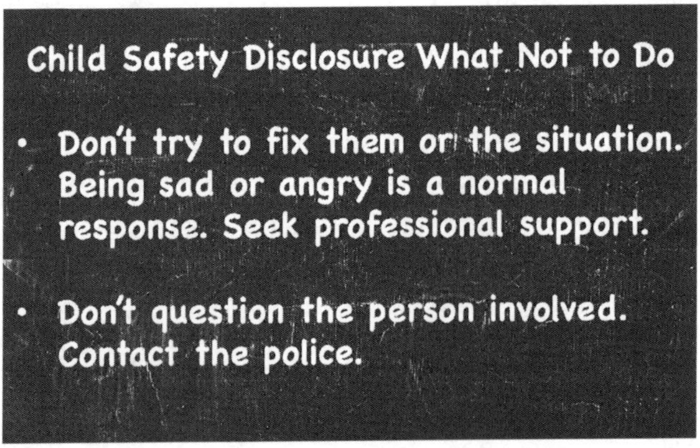

> ## Child Safety Disclosure What Not to Do
>
> - Don't try to fix them or the situation. Being sad or angry is a normal response. Seek professional support.
>
> - Don't question the person involved. Contact the police.

Figure 3.4   What not to do[8].

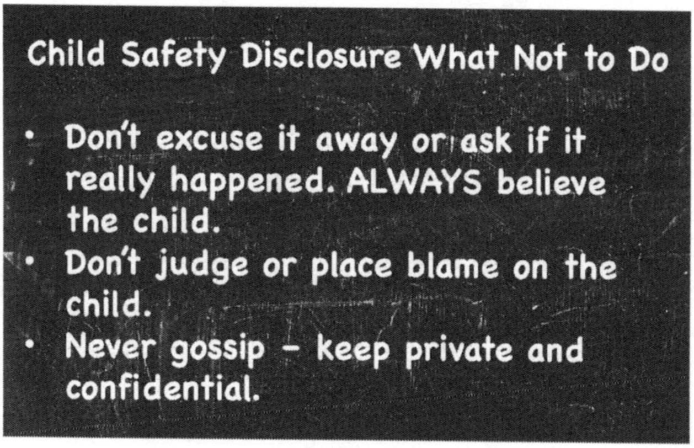

Child Safety Disclosure What Not to Do

- Don't excuse it away or ask if it really happened. ALWAYS believe the child.
- Don't judge or place blame on the child.
- Never gossip – keep private and confidential.

Figure 3.5 What not to do (continued[9]).

autism spectrum disorder, the list of rules for tricky people may be used in that environment and is also applicable to all children.

The following are some rules for children with intellectual disability and autism spectrum disorder about "tricky people" (summarized from the Safely Ever After Program):

1. People must ask your permission before touching you in any way. This includes hugs and kisses. Your child does not need to apologize or have a reason for not wanting to hug or kiss someone.
2. Everybody has private body parts that must be covered when you are in public.
3. Make sure your child knows their full name along with the full names of their parents/carers.
4. Your child should never go anywhere with someone they don't know or take anything from someone they don't know. Things like ordering takeaway or asking for something at the shops are different because you are requesting help from someone.
5. Your child should always check with their parents/carers before:
   - changing plans without prior notice
   - getting into a car (even if the driver is someone they know)
   - accepting gifts (gifts should never be a secret)

6. Your child does not have to be polite if they feel scared or uncomfortable.

7. Your child is allowed to say or sign NO to adults and other children.

8. No, Go, Tell: Teach your child to say or sign no, and to then go and tell their safe people.

9. Your child should never be asked to keep a secret. Secrets can make us feel scared and uneasy. If there is information your child must keep to themselves, like a party or a gift for another person, reframe this as a SURPRISE! No adult should ever ask your child to keep a secret.[10]

Remember that safe touch is good! Don't forget to give your child examples of positive safe touch. For example, high fives with friends or a hug from a safe adult that makes you both feel good.

A good exercise may be to have students identify who are their safe people. Create a classroom list of safe people, encouraging students to share their ideas of safe people. Be sure to continue supporting good and positive examples.

Teachers can be a safe, trusted adult for children.

Darkness to Light, an organization with the mission to end child sexual abuse, provides methods for you to be a safe adult — so that a child can trust you to be there for them when they need help.

• Set and maintain clear, protective boundaries. This is a crucial step because boundaries help us to respect each other and feel respected. A child who knows that you will respect (and even protect!) their boundaries will have an easier time trusting you to take care of them.

• Develop protective bonds. Really listen when the child talks to you. Show them that you're interested in their opinion; involve them in conversations and show them that their input is valuable.

• Talk openly and honestly about child sexual abuse. When we talk to children in age-appropriate ways about our bodies, sex, and boundaries, children understand what healthy relationships look like. It also teaches them that they have the right to say "no."[11]

The National Center for Missing and Exploited Children® suggests using the following language when talking to your child about abduction prevention:

DON'T SAY

> Never talk to strangers.

DO SAY

> You should not approach just anyone. If you need help, look for a uniformed police officer, a store clerk with a nametag, or a parent with children.

DON'T SAY

> Stay away from people you don't know.

DO SAY

> It is important to get my permission before going anywhere with anyone.

DON'T SAY

> You can tell someone is bad just by looking at them.

DO SAY

> Pay attention to what people do. Tell me right away if anyone asks you to keep a secret, makes you feel uncomfortable, or tries to get you to go somewhere with them.[12]

These tips and the language may be adapted and used in various discussions about tricky and safe/unsafe people.

Gaming and Avatars — What's Safe, What's Not

Children at this age level now have some experience with online games and may even have an avatar. An avatar is a representation of a person. It may be an expression of their interest and likes. It may even be how they see themselves, especially in online activities. It Is not just a distortion of an actual photo of the individual but may also be a character or object (Figure 3.6).

The Pew Research Center's 2020 study shows that 80% of parents with a child age 5–11 say their child uses or interacts with a tablet

Figure 3.6   An avatar.[13]

computer, compared with 64% of parents with a child age 3–4 who do this and 35% with a child or a child age two or younger (Figure 3.7).[14]

There are many online sites to create an avatar including most gaming sites. A source for understanding the media available, at an age-appropriate level, is the non-profit CommonSense.org. Believing

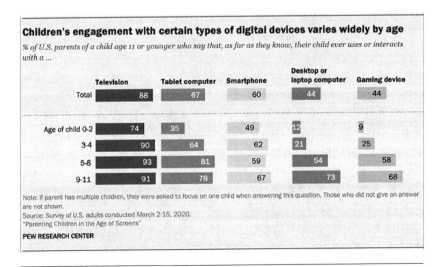

Figure 3.7   Pew research — children's engagement with digital devices by age.[15]

that most technologies weren't built with kids in mind, Common Sense is on a mission to change that. Common Sense rates movies, TV shows, apps, podcasts, books, and more so families can feel good about the entertainment choices they make for their kids. Offering a library of independent age-based ratings and reviews helping families make choices about what to play, read, and watch.[16]

The plus to an avatar is that the child's picture is not posted making it unavailable for viewing, manipulating, and even face recognition searches, providing an additional layer of safety. This is also the negative risk associated with an avatar. Like a mask, bad actors may hide behind an avatar, keeping others from seeing who the person is (Figure 3.8).

Hiding behind an avatar is an ideal way for an adult to gain access to various gaming sites for children. The sole purpose is to "groom" the child for potential nefarious activity such as an in-person meeting often with the goal of committing sexual abuse. Many apps will have a chat feature thought to be only for gamers within the app, appearing as a safe place to chat with others.

As an exercise, you can talk with the students about masks and why we use masks.

Ask students how they feel when they are talking with someone wearing a mask and they are unable to see the person.

Figure 3.8 Anonymous computer hacker.[17]

How they feel when they are wearing a mask? Are they more secure or bold because they are hidden?

Many of the apps become a backdoor to Internet access. School devices or devices students bring from home into school or after school activities may have websites blocked or the Internet disabled. An app may seem child-safe yet allows access to other sites and chats that are not. Once the bad actors behind the avatar befriend the child, they will invite the child to move to a different social media site. The chat feature within the app may circumvent the assigned controls. Monitoring is crucial.

Bill Sweeney, with the FBI office in New York City, cautions parents and guardians. As soon as the "instant message pops up on your child's browser or on their smartphone, there could be a sexual predator on the other end. It takes only a few days — sometimes just hours — to groom your child into sending compromising photos to the person on the other end of that chat. Then the threats begin. That predator starts to ask for photos of your child — something they would be ashamed of if anyone, including you, found out. Then they threaten to post them online if they don't keep sending more and more graphic photos."[18]

Other Ways Apps Can Compromise Young Users

App developers and the companies marketing them want to sell more products, services, and other apps. Pop-up blockers are great — especially when they are used — but are not fail-safe. Many sites will ask for "demographic" information during set-up or in the creation of the avatar. Site policies vary and for teachers, guardians, and parents this should be a supervised process.

According to the Washington Post article *Your Kids' Apps are Spying on Them* children's privacy "deserves special attention because kids' data can be misused in some uniquely harmful ways. Research suggests many children can't distinguish ads from content. This is why kids are at the center of one of America's few privacy laws, the 1998 Children's Online Privacy Protection Act, or COPPA." Although this act has been on record since 1998, a big loophole remains. Tech companies and app designers argue that they do not "actually" know they are taking data from kids.[19]

As we read in Chapter 2, apps create an ideal opportunity to provide personally identifiable information (PII). A series of questions may be asked during the set-up process. Frequently this step is not required to gain access to the app and may often be skipped. There are also many engaging and fun apps and games accessible without the Internet. Although the questions may seem innocent, the PII provided may be used by groomers to lure children to other sites. When the questions refer to other family members, pets, and pet names, or where you live or go to school, hackers may use this information to gain access to other family members. Parental and security controls are critical to keeping young users safe from tricky and unsafe people, whether the school provides devices or students bring their own device to school.

An Additional Level of Safety — A Family Safe Word

One way to ensure that someone is not a dangerous or tricky/unsafe person is to create a family safe word. This is a word or a couple of words known to only the immediate family. It is not shared with other relatives, friends, or neighbors. It is easy enough for children even at the youngest of ages to remember.

How and When to Use the Family Safe Word

The safe word may be used by the child in any environment where a child feels unsafe or uncomfortable with their surroundings. It makes them feel that they have control in many situations. An example provided by The Pragmatic Parent places the child at a friend's BBQ where the child is with an adult who is making inappropriate conversations and requests of your child, your child can come to you, and tell you the safe word regardless of whether you are alone or standing around in a group with your friends — you will know exactly what they are disclosing and can respond immediately.[20]

A safe word can be used anywhere — you may even create a class-room-safe word. This allows you and the child to have a code for all situations. The importance of using the safe word must be strongly conveyed — think of the child who cried wolf one too many times. Examples of the safe word may be used at a school sporting event, after-school activities, and any school-sponsored event where the child feels uncomfortable by an adult or even another student.

Summary

By raising awareness among children, they can learn to become responsible for understanding that they can say "No" to an adult if they are uncomfortable or it feels "icky" and that they have the tools to identify who the helpers are and who may be playing a trick on them. It is not too early to create a sense of ownership of the body and personal safety.

Chapter 4 presents the emerging topic of "Exploring Cyber Safety Through Stewardship."

LESSON PLANS

Grade 3

TOC Title: G3 Safe and Unsafe People
Lesson Title: Safe and Unsafe People
Grade Level: 3
Duration: 45 minutes

Objective:

- The students will be able to identify safe and unsafe people.

Suggested Materials:

- Cards with pictures or descriptions of behaviors that help identify safe and unsafe people, made ahead of time. Some suggestions for cards follow this lesson plan.

Procedure

Introduction (10 minutes):

a. Class discussion about the Internet, emphasizing its positive aspects (learning, entertainment) and potential dangers.
b. Introduce the concept of safe and unsafe people online. Discuss what makes someone safe or unsafe. Explain that safe people make us feel comfortable, respected, and cared for, whereas unsafe people will make us feel uneasy, uncomfortable, or scared.
c. Discuss the importance of recognizing and choosing safe people in our lives.
d. Refer to chapter content as needed.

Activity 1 (10–15 minutes):

Sorting Safe and Unsafe Characteristics:

a. Class may be divided into small groups (3–4).
b. Provide each group with a set of cards depicting different characteristics or behaviors.

- Some examples of safe characteristics include smiling, sharing, listening, and being helpful.

- Some examples of unsafe characteristics include frowning, teasing, ignoring others, being mean.

c. The students sort the cards into "Safe" and "Unsafe" piles. The students should talk within their groups about their choices (why this is a safe/unsafe characteristic).

Activity 2 (10–15 minutes):

Class Discussion:

a. Whole group: ask students to describe how and why they categorized certain behaviors as safe or unsafe.
b. Emphasize that personal safety includes physical safety as well as emotional well-being.

Conclusion (5 minutes):

a. Summarize key points of the lesson.
b. Point out the importance of talking to a trusted adult if they ever feel unsure or uncomfortable with someone.

Assessment:

a. Observe students' participation in the activity and the class discussion to assess their understanding of safe and unsafe people.

Homework (optional):

a. Ask students to draw a picture of a safe person at home and discuss the topic of safe and unsafe people with their family.

Suggestions for Characteristics for Cards:

Safe Person Characteristics:

a. Kind
b. Caring
c. Good listener
d. Trustworthy
e. Respectful
f. Helpful
g. Reliable
h. Non-judgmental
i. Compassionate
j. Responsible
k. Open-minded
l. Generous

Unsafe Person Characteristics:

a. Bullying
b. Manipulative behavior
c. Untrustworthy
d. Self-centered
e. Intolerant
f. Lacking empathy
g. Bad influence
h. Disrespectful
i. Defensive
j. Controlling
k. Secretive
l. Gossips

TOC Title: G3 Online Gaming and Avatar Safety
  Lesson Title: Online Gaming and Avatar Safety
  Grade Level: 3
  Duration: 60 minutes (may be split into 35- and 25-minute sessions)

Objective:

- The students will understand the importance of online gaming safety.
- The students will be able to identify the pros and cons of using avatars in online games.
- The students will develop strategies for staying safe while gaming online.

Suggested Materials:

- Paper.
- Markers, crayons, or colored pencils.
- Printed handouts of sample avatars (prepare ahead of time) from a link such as free avatar images no copyright - Search (bing.com) (https://www.bing.com/search?q=free+avatar+images+no+copyright&FORM=QSRE2) or a similar site.
- Chart paper or whiteboard.
- Drawing materials.

Procedure

Introduction and Discussion (15 minutes):

a. Begin by asking the students if they have ever played games online. Invite them to share their experiences.
b. Discuss the importance of online gaming safety.
c. Ask questions such as:

- Why is it important to be careful when playing games online?
- What are some potential dangers of interacting with others online?
- How can we stay safe while gaming online?

d. List student responses on the whiteboard.

e Emphasize the following safety tips:

- Never share personal information online, including name, address, phone number, or school.
- Use a nickname or username instead of real names.
- Be careful when chatting with other players and always report any inappropriate behavior to a trusted adult.
- Only play games that are age appropriate.
- Limit screen time and take breaks while gaming.

f. Discuss what avatars are and why they are used online (to represent a player. This can be a harmless form of self-expression or conceal the identity of a dangerous person).

g. Tell students that they will be designing their own avatars and safety posters to promote online gaming safety.

h. Clarify that an avatar should not have information that can be connected to the actual person.

Avatar Creation (20 minutes):

a. Distribute printed handouts of sample avatars to each student.

b. Alternatively, students could be allowed to create their own avatars from scratch if appropriate.

c. Instruct students to color and customize their avatars using markers, crayons, or colored pencils.

d. Encourage creativity and remind students to think about how they want to represent themselves online through their avatars.

Stop here if doing two lessons

If doing as a second lesson, briefly review discussion from Lesson 1.

Safety Poster Design (20 minutes):

a. Divide the class into small groups (2–4 students suggested).

b. Provide each group with large paper.

c. Students create a safety poster that promotes one or more online gaming safety tips discussed in the lesson.

d. Encourage students to use their avatars in the poster design to illustrate safe online gaming practices.

e. Remind students to include clear and easy-to-understand messages on their posters.

f. Once the posters are complete, have each group present their poster to the class.

g. Allow students to explain the safety tips depicted on their posters and how they used their avatars to promote online gaming safety.

Conclusion (5 minutes):

a. Conclude the activity with a brief reflection.

b. Ask students to share one thing they learned about online gaming safety or avatars during the activity.

c. Emphasize the importance of staying safe while gaming online and remind students to apply the safety tips in their own online gaming experiences.

Assessment:

a. Observation of student engagement and participation during the avatar creation and poster design activities.

b. Assess students' understanding through their contributions during the poster presentations and reflections on what they learned.

Homework (optional):

a. Encourage students to share what they learned about cyber safety with their families.

Grade 4

TOC Title: G4 Safe/Unsafe People
  Lesson Title: Safe/Unsafe People
  Grade Level: 4
  Duration: 40 minutes

Objective:

- The students will understand the difference between safe and unsafe people.
- The students will differentiate between safe and unsafe online activities.
- The students will understand and use strategies for staying safe online.

Suggested Materials:

- Whiteboard and markers.
- Printed handouts of safe and unsafe online activities list (suggestions included after lesson).
- Scenario cards for assessment, which are provided at the end of this lesson plan.

Procedure

Introduction (5 minutes):

a. Lead a class discussion about online safety. Encourage students to share relevant experiences and knowledge about using the Internet.
b. Explain that, just like in the real world, there are safe and unsafe activities and people online.

Activity 1 (15 minutes):

Safe versus Unsafe:

a. Class should be in groups of 3–4 students.
b. Provide each group with a random list of safe and unsafe online activities taken from the list provided below. Suggestion: print several copies of both lists, cut out and mix them up before handing them out.

c. Have students discuss and categorize each activity in their set as safe or unsafe. They should also discuss reasons for their choices. The teacher monitors and guides making sure activities are sorted correctly. If one is sorted incorrectly, offer discussion or ask questions that help the group understand where the activity should be placed.

d. After 10 minutes, come back together as a class to discuss the findings. Write down common characteristics of safe and unsafe online activities on the whiteboard.

Activity 2 (15 minutes):

Scenario-Based Activity/Assessment:

a. Distribute scenario cards to each student.

b. Students read the scenarios and write a brief explanation of how they would handle the scenario.

c. Collect when done.

d. Review the scenarios as a class and discuss the safe and unsafe aspects of each situation.

Conclusion (5 minutes):

a. Summarize the key points discussed during the lesson.

b. Answer any last questions students have about the topic.

Assessment:

a. See Activity 2 above.

Materials for Activity and Assessment

Activity 1:

Safe and Unsafe Online Activities:

Safe Online Activities:

   a. Playing educational games on trusted websites.

   b. Communicating with known friends and family members through email or messaging apps with parental supervision.

   c. Researching school projects using reliable search engines like Google SafeSearch.

   d. Watching age-appropriate videos or TV shows on approved streaming platforms.

   e. Collaborating on group projects using secure online platforms provided by the school.

   f. Creating an avatar you like as a representation of yourself that does not reveal any personal information an unsafe person could use to find you.

   g. Using password-protected accounts for online activities.

   h. Reporting any suspicious or inappropriate content to a trusted adult.

   i. Using privacy settings to control who can see personal information on social media platforms (if age-appropriate and with parental guidance).

   j. Following Internet safety rules provided by parents or guardians.

Unsafe Online Activities:

   a. Sharing personal information like your full name, address, phone number, or school name with strangers online.

   b. Clicking on pop-up ads or links in emails from unknown senders.

   c. Engaging in cyberbullying by sending mean or hurtful messages to others.

   d. Downloading files or software from untrustworthy sources.

e. Visiting websites with inappropriate content, such as violence, explicit language, or adult material.

f. Accepting friend requests or messages from strangers on social media without verifying their identity.

g. Sharing passwords or login information with anyone other than parents or guardians.

h. Participating in online challenges or in dares that could be dangerous or harmful.

i. Posting or sharing photos or videos without permission from those involved.

j. Creating an avatar using your school mascot, your first name and last initial.

Note: These items can be altered or expanded if needed to better match the specific needs of your students.

Scenarios for Activity/Assessment

Scenario 1:

You receive a friend request on social media from someone you don't know. The two of you have no mutual friends. They message you and ask for your address because they want to send you a gift card. What do you do?

Scenario 2:

While playing an online game, another player starts using mean language and insults you. They ask for your personal information, like your age and where you live. What should you do?

Scenario 3:

You receive an email with a link that says you've won a free iPad. It asks you to click the link to claim your prize. What should you do?

Scenario 4:

While chatting with someone online who claims to be a kid of your age, they ask you to share your phone number and suggest meeting up in person. What do you do?

Scenario 5:

You are searching for information for a school project and come across a website with inappropriate pictures and words. What should you do next?

Scenario 6:

You receive a message from someone claiming to be your friend's parent. They ask for your home address because they want to send you a gift for being such a good friend to their child. What should you do?

Scenario 7:

In a chat room, someone asks for your real name and the name of your school. They say they want to become friends with you. What do you do?

Scenario 8:

While playing a game online, someone asks you to share your password with them so they can give you extra points or special items. What should you do?

Scenario 9:

You find a website that promises to give you free movies to download, but it asks for your credit card information. What should you do?

Scenario 10:

You receive a message from someone online who says they know you from school. They ask for your home address. You do not recognize them. What do you do?

TOC Title: G4 Online Gaming and Avatar Safety
   Lesson Title: Online Gaming and Avatar Safety
   Grade Level: 4
   Duration: 55 minutes

Objective:

- The students will understand the importance of online gaming safety.
- The students will be able to identify both positive and negative aspects of avatars.

Suggested Materials:

- Whiteboard and markers.
- Paper and coloring supplies.
- Printouts of online gaming safety tips or poster, which may be found at Online Safety Leaflets & resources - Internet Matters (https://www.internetmatters.org/resources/esafety-leaflets-resources/), or a similar site.
- Poster board or large paper for activity.

Procedure

Introduction (5 minutes):

a. Begin by asking students if they have played games online. Discuss what types of games they play and if they use avatars.
b. Explain that today's lesson will focus on online gaming safety and avatars.

Discussion on Online Gaming Safety (10 minutes):

a. Discuss online gaming safety. Point out that playing games online can be fun, but it is important to always stay safe.
b. Have students brainstorm safety tips for online gaming. If students need suggestions to get started, some examples are never share personal information, keep gaming accounts private, and always be cautious when chatting.
c. Write these tips on the whiteboard or chart paper and post where students can easily see it.

Activity 1 (15 minutes):

Pros and Cons of Avatars:

a. Define avatar as a digital representation of a person in a game or virtual world.
b. Divide the class into two groups: "Pros" and "Cons."
c. Have each group brainstorm the positive and negative aspects of using avatars in online games. Examples: Pros — Identity exploration, personalization, social interaction. Cons — anonymity (adult masquerading as a child), inappropriate or offensive behavior, and avatars that perpetuate stereotypes.
d. Each group will share their ideas with the class. Write them on the whiteboard.

Activity 2 (20 minutes):

Design Your Avatar:

a. Tell students that they are going to design their own avatars.
b. Provide paper and coloring supplies.
c. Encourage students to think about how they want their avatars to represent them.
d. Once finished designing, have students share their avatars with the class and explain why they chose certain features.

Conclusion (5 minutes):

a. Review the key points of the lesson: online gaming safety and the positive and negative aspects of avatars.
b. Remind students to always be safe when gaming online and to make smart choices with their avatars.

Assessment:

a. The teacher should take an informal assessment during activities to gauge student understanding and participation in the activities.

Homework (optional):

a. This is not an assignment, per se, but consider printing it out and sending it home for students to share with their families.

Tips for Online Gaming

1. Never share personal information such as your full name, address, phone number, school name, or passwords with anyone online even if they claim to be a friend.

2. Use a safe username. Choose a username that does not reveal personal information or attract unwanted attention. Avoid using your real name or any identifying information in your username.

3. Keep your gaming accounts private to control who can interact with you online.

4. Be cautious when chatting. Never share personal information in chat and immediately report any inappropriate behavior or messages to a trusted adult.

5. Play with friends you know in real life rather than with strangers.

6. Take regular breaks to rest your eyes, stretch, and do other activities. Balance game time with other activities and responsibilities.

7. Always be respectful and kind. Bullying, teasing, or being rude to other players is never acceptable behavior.

8. Trust your instincts. Immediately leave a game or chat if you feel uncomfortable or threatened in any way. Remember that it is okay to seek help from a trusted adult if you encounter any problems while gaming online.

Notes

1 Rogers-Nelson, K. January 18, 2018, "'Stranger Danger' is Over — Here's What Parents are Teaching Their Kids Instead," https://www.sheknows.com/parenting/articles/1137790/stranger-danger-is-over-tricky-people/, retrieved February 22, 2023.

2 Centers for Disease Control and Prevention, Fast Facts: Preventing Child Abuse, Fast Facts: Preventing Child Sexual Abuse |Violence Prevention|Injury Center|CDC, Fast Facts: Preventing Child Abuse & Neglect |Violence Prevention|Injury Center|CDC, retrieved January 16, 2024.

3 Shutterstock image, February 29, 2024, used under CRC License.

4 Showfety, S. July 20, 2022, "The Best Way to Teach your Kids to Recognize a Safe Adult," https://lifehacker.com/the-best-way-to-teach-your-kids-to-recognize-a-safe-adu-1849194970, retrieved November 27, 2023.

5 Kids First, Inc. https://www.kidsfirstinc.org/how-to-talk-to-young-children-about-body-safety/, retrieved November 27, 2023.

6 Figure 3.2 adapted by authors from The Gentle Counsellor, https://thegentlecounsellor.com/talking-to-your-child-about-safety-strangers-tricky-people/#:~:text=Tip%203%3A%20Teach%20your%20child,someone%20known%20to%20the%20child.

7 Figure 3.3 adapted by authors from The Gentle Counsellor, https://thegentlecounsellor.com/talking-to-your-child-about-safety-strangers-tricky-people/#:~:text=Tip%203%3A%20Teach%20your%20child,someone%20known%20to%20the%20child.

8 Figure 3.4 adapted by authors from The Gentle Counsellor, https://thegentlecounsellor.com/talking-to-your-child-about-safety-strangers-tricky-people/#:~:text=Tip%203%3A%20Teach%20your%20child,someone%20known%20to%20the%20child.

9 Figure 3.5 adapted by authors from The Gentle Counsellor, https://thegentlecounsellor.com/talking-to-your-child-about-safety-strangers-tricky-people/#:~:text=Tip%203%3A%20Teach%20your%20child,someone%20known%20to%20the%20child.

10 Planet Puberty, Identifying safe people - Planet Puberty, retrieved January 16, 2024.

11 Darkness to Light, April 3, 2020, "How to be a Safe Adult", https://www.d2l.org/how-to-be-a-safe-adult/, retrieved November 27, 2023.

12 "Rethinking Stranger Danger", Kids Smartz, The National Center for Missing and Exploited Children®, https://www.missingkids.org/education/kidsmartz, retrieved December 1, 2023.

13 Shutterstock image, March 9, 2024, used under CRC License.

14 Figure 3.7 Adapted by authors from Pew Research Center, July 28, 2020, "Parenting Children in the Age of Screens," Children's engagement with digital devices, screen time, page 13, www.pewresearch.org/internet/2020/07/28/parenting-children-in-the-age-of-screens, retrieved February 22, 2024.

15 Pew Research Center, July 28, 2020, "Parenting Children in the Age of Screens," Children's engagement with digital devices, screen time | www.pewresearch.org/internet/2020/07/28/parenting-children-in-the-age-of-screens, retrieved February 22, 2024.

16 https://CommonSense.org, retrieved February 22, 2024.

17 Shutterstock image, March 9, 2024, used under CRC License.

18 Sweeney, B., 2024. "It's Not a Game: FBI New York PSA," It's Not a Game: FBI New York PSA — FBI, retrieved February 22, 2024.

19 Fowler, G. A. Washington Post, June 9, 2022, "Your Kids' Apps are Spying on Them" Apps violate kids' privacy on a massive scale - The Washington Post, retrieved February 22, 2024.

20 The Pragmatic Parent, "Why Every Family Needs a Safe Word." https://www.thepragmaticparent.com/safeword/#:~:text=A%20safe%20 word%20is%20a,as%20puppy%2C%20soccer%20or%20milk, retrieved February 24, 2024.

4

EXPLORING CYBER SAFETY THROUGH STEWARDSHIP

This chapter emphasizes the concept of stewardship in both physical and virtual environments. It underlines the importance of responsible online behavior, care for digital and physical possessions, and environmental stewardship. This chapter integrates science and engineering practices with environmental education, highlighting the role of educators in fostering stewardship through student-centered projects. It also discusses the significance of digital stewardship, managing digital assets responsibly, and ensuring safe, ethical use of digital information. Key principles of good digital stewardship are outlined, including respect for digital devices, responsible online behavior, protection of personal information, and the development of positive digital habits.

Introduction

What has been discussed in previous chapters and learned through in (and out of) class activities may now culminate in focusing the student's attention on acting responsibly when engaging in online activities, on acting safely with their and other people's possessions, with other people, and when interacting with the environment in which they live, learn and play.

Why is addressing the concept of stewardship important at this age level? First, let us start by defining stewardship and taking a look at the evolution of the concept of stewardship over time.

The Merriam-Webster Dictionary defines stewardship as the responsible overseeing and protection of something considered worth caring for and preserving. Expanding upon the initial definition, stewardship may be viewed as the careful and responsible management or oversight of something entrusted to one's care, intending to

DOI: 10.1201/9781003466338-4

preserve, protect, and enhance its value for the benefit of others or for future generations.

Good stewardship is about taking care of things placed in our care to enjoy them now, and for future generations to be able to enjoy them later. Building the foundation for stewardship early will create a sense of responsibility in young learners.... the eventual caretakers of our planet Earth.

Stewardship or the responsible management and care of resources or assets that have been entrusted to an individual or organization has been in place for centuries. The meaning of the word "stewardship" has evolved, with its usage and connotations shifting across different historical periods. The following is a summary of the historical evolution of the meaning of stewardship, with specific periods assigned for each evolution:

The Meaning of Stewardship — Historical Evolution

Ancient Times   The concept of stewardship has roots in ancient civilizations, where stewards were appointed to manage estates, land, or household affairs on behalf of their rulers or landlords. This role was crucial in ensuring the prosperity and order of estates and often included managing agricultural activities, collecting rents, and overseeing servants.

Medieval Period (5th to 15th Century)   In medieval Europe, the role of the steward became more formalized within the feudal system. Stewardship primarily referred to the management and administration of large estates or households owned by nobility or the church. The steward was a high-ranking official, their duties extended to judicial roles, presiding over estate courts, and responsibility for overseeing domestic affairs, finances, and various properties.

Early Modern Period (16th to 18th Century) — Expansion of Stewardship Role   In this period, the concept of stewardship began to extend beyond the management of estates and households. It was used in a broader sense to refer to the responsible management and care of resources or assets entrusted to someone, often with a moral or ethical dimension. Stewards played key roles in managing the affairs of states and were often seen as trustees or guardians of public and private wealth.

19th Century — Industrial Revolution — Shift Toward Corporate Stewardship During the 19th century, stewardship further expanded to encompass the responsible management and preservation of natural resources, such as land, forests, and wildlife. This period saw the concept of stewardship evolve to include corporate governance, with stewards (or directors) responsible for managing the resources of a company in the best interest of its shareholders. This period also saw the emergence of conservation movements and a growing awareness of the need for environmental stewardship.

20th Century — Environmental Stewardship In the 20th century, the concept of stewardship became more widely applied to various domains, including corporate governance, public policy, and social responsibility. Stewardship came to encompass the responsible use and protection of the natural environment through conservation and sustainable practices. The term "corporate stewardship" emerged, referring to the ethical and responsible management of a company's resources, assets, and stakeholder interests. Stewardship principles were also applied more broadly in the management of non-profit organizations and public sector entities, emphasizing accountability, transparency, and the ethical use of resources to serve the community or public interest.

In contemporary usage, stewardship has taken on an even broader meaning, encompassing the responsible care and management of a wide range of resources, including the environment, cultural heritage, communities, and personal assets or responsibilities. The concept has been extended to areas such as sustainability, social impact, and inter-generational equity.

Digital Stewardship (Mid-21st Century and Beyond) The rise of digital information and technology has led to the concept of digital steward-ship, which involves the management and preservation of digital data and resources for future accessibility and use.

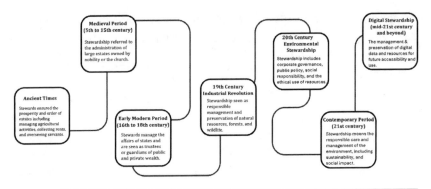

Figure 4.1   The meaning of stewardship — historical evolution.[1]

From its origins in managing estates to its current application in environmental protection, corporate governance, global sustainability efforts, and most recently, digital stewardship the meaning of stewardship has evolved; however, its essence of responsible oversight, care, and preservation of valued resources or assets has remained consistent throughout its historical development. Each period reflects the changing societal values and challenges of the times, demonstrating stewardship's adaptability and enduring relevance (see Figure 4.1).

The concept of stewardship is often applied in various contexts, including environmental stewardship, financial stewardship, organizational stewardship, and most recently digital stewardship.

In the context of the environment, stewardship involves taking care of natural resources such as land, water, and biodiversity to ensure their sustainable use and conservation for future generations. This may include practices such as conservation, sustainable agriculture, and pollution reduction.

Financial stewardship pertains to the responsible management of financial resources, whether it be personal finances, corporate finances, or funds belonging to an organization or institution. It involves making sound financial decisions, managing budgets effectively, and ensuring transparency and accountability in financial transactions.

Within organizations, stewardship refers to the responsible management of the organization's resources, including its human capital, physical assets, and reputation. This involves promoting ethical behavior, fostering a culture of accountability, and ensuring that resources are used efficiently to achieve the organization's mission and goals.

At this age, students are learning to navigate responsibility and understand how to act in a responsible role. Responsibility extends to themselves, their classmates, the adults around them, and their community. Parenting Montana cites research confirming that children are developing cause-and-effect thinking. This directly impacts their capacity to take responsibility for their actions. Once they understand how their actions and decisions affect not only themselves but also those around them, they will approach even the smallest things they do in their day with a sense of responsibility and pride.

Parenting Montana recommends considering the five-step process to help foster responsibility and make good choices.

1. Get children thinking by getting their input. Ask open-ended questions to prompt thinking.
2. Teach new skills by interactive modeling. Learning new skills and behaviors requires modeling, practice, support, and recognition.
3. Practice to grow skills and develop habits. Daily routines are opportunities to practice.
4. Support the child's development and success. Offer support when needed by reteaching, monitoring, coaching, and, when appropriate, applying logical consequences.
5. Recognize effort and quality to foster motivation. Your praise and encouragement are their sweetest reward.[2]

How Do We Foster Stewardship among Children?

Children have "grown-up" with technology. It is embedded in their daily routine. They are neither hesitant nor afraid of where it may take them. As teachers and adults, we have a responsibility to help them be good stewards of what has been entrusted to them.

Because science and engineering practices are often underemphasized in K—12 science education, The National Science and Teaching Association's Environmental Education Professional Development Institute (EEPDI) emphasized pathways to integrating next-generation practices with environmental stewardship for informed, responsible action on behalf of the environment and future generations.[3]

The EEPDI Focus emphasizes the integration of *Next Generation Science Standards* (NGSS) practices through student-centered stewardship projects, which offer an excellent means to get students involved in science, increase their critical thinking, and motivate action on environmental issues that are meaningful to the students. One goal was to engage students in "real" local environmental issues by getting them outdoors working in collaborative groups.

NSTA concludes that you too, like the teachers involved in the Environmental Education Professional Development Institute, can incorporate the *NGSS* Science Practices into your student-driven stewardship projects. These activities can be adapted to student abilities. The project teachers found it helpful to also seek cross-disciplinary (e.g., math teachers) collaboration at their school, as well as recruit community expertise outside of their schools.

Science and engineering practices were integrated as students asked questions, planned and carried out their investigation, analyzed and interpreted data, and communicated information. Since repeated exposure is required over many years to deeply understand the disciplinary core ideas (DCIs) and crosscutting concepts (CCCs) and to build students' proficiency with the practices, it is proposed that repeated, varied stewardship projects provide an excellent opportunity for three-dimensional learning.

According to Lesley University,[4] giving children opportunities to be connected to the natural world and ensuring children become ecologically literate citizens who have a sense of ownership and stewardship of the earth. Environmental educators in particular have been instrumental in creating awareness, programs, and opportunities for children of all ages to connect with the natural world, knowing that our future depends on ecologically literate citizens who will have a sense of ownership and stewardship of the earth. A No Child Left Inside movement has spawned as a result.

Here are seven specific suggestions from Lesley University's science faculty, which you may use or adapt to achieve the right age-appropriate level.

1. Allow children unfettered time in the natural world.
   This means not organized sports or adult-directed activities, but lengthy time to explore, play, and invent. It might

take the form of building a fort from found objects, damming a stream, or collecting natural objects like shells, rocks, or acorns. These moments in nature are recalled later, as adults, in vivid descriptions.

2. Be a Mentor.

Rachel Carson, in her 1956 article entitled "A Sense of Wonder," implored adults to find one child to mentor in the workings of the natural world. Whether parent, grandparent, or other relative or friend, most adult conservationists can point to those people in their lives who had significant influence on them.

Create opportunities for children to have experiences with the more-than-human world. Volunteering at a wildlife rehabilitation center, snorkeling on a vacation, or simply walking in nature and having surprise contact with local species are all important ways children can come in contact with the species that share their home. Speak of them as friends and talk about how they are connected to humans.

3. Study the local bioregion.

Guide children to understand the area they live in, for instance, where their water comes from and where it goes once it leaves their home or school; what plants are native or non-native, wild or cultivated; what animals share their home with them; how people make a living from the earth's resources; and what natural wonders — ponds or streams, marshes, hills, and so on — are nearby.

4. Engage children in real-life actions.

Whether it is planting trees, creating a garden, pulling invasive species, or picking up garbage — begin the stewardship mindset.

5. Enroll your child in a real outdoor camping program.

In addition to a physical fitness camp, technology camp, or sports camp, find a camp where children sing by the campfire, sleep out under the stars, learn to make bows and arrows, learn to steer a canoe, or learn to use a bow drill to make a fire.

6. Get to know your state or national parks.

If you are lucky enough to live near a protected area, visit. These parks are protected for a reason and offer wonderful opportunities for hiking exploring, and experiencing the sounds, smells, and excitement of unfamiliar natural environments.

Earth.org reinforces connecting students with nature and the outdoors. "Parents and teachers can help students understand their role as environmental stewards by encouraging student outdoor learning programs and supporting young folks who engage in student activism."[5] If your school or nearby park has natural playgrounds, like those built from sustainable materials and found objects, Earth.org indicates this is a perfect place and opportunity to discuss environmental protection and the importance of stewardship over the Earth's resources.

Screen time is increasing at all ages, and the use of social media is growing among 8- to 12-year-olds. Thirty-eight percent of tweens have used social media (up from 31% in 2019), and nearly one in five (18%) now say they use social media "every day" (up 5% points since 2019).

The time spent using social media is also up to 8 minutes a day among this age group (from 10 to 18 minutes a day, on average). Social media was defined in the survey as being sites such as Snapchat, Instagram, Discord, Reddit, or Facebook; platforms such as YouTube, TikTok, and Twitch were considered online video sites.

The top five social media sites tweens have ever used are Snapchat (13%), Instagram (10%), Facebook (8%), Discord (5%), and Pinterest (4%) (see Figure 4.2).

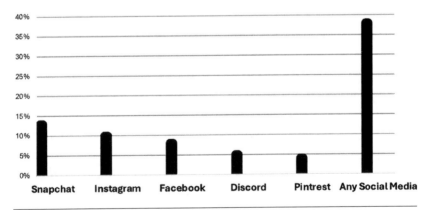

Figure 4.2   Top social media sites among tweens, 2021.[6]

What Does Stewardship Mean to the Teacher's Role?

In the article "The Importance of Stewardship in Leadership" from the Graduate Programs for Educators, the concept of the teacher as a wise steward of leadership in the classroom is examined.[7] Burress believes these are the expected areas of stewardship when serving as an educator leader. Each of these areas is an important factor in becoming an effective steward of educational leadership.

Stewardship of Students

First and foremost, the most important thing we are entrusted with is our students. What does it mean to be a steward for students? Our communities trust us with their students for 180 days a year. However, more importantly, parents trust us with their children! We need to always ask ourselves the following daily questions:

1. Is what we are doing in the best interest of students?
2. Would you entrust your child with our instructors?
3. Are we making an impact on these children's lives?

As educators, we should never take the responsibility of stewardship of other people's children lightly. Education is a calling; being a wise steward of students is a responsibility that is accompanied by passion, enthusiasm, and joy.

Stewardship of Influence

Educators have a remarkable amount of influence over their students, and educational leaders have the same influence over their staff. We are entrusted to use that influence in such a way that our instructional teams can help our students become productive members of society. We have all heard stories of educators telling students that they are not college material or shouldn't do something because they don't have the right skill set.

Words matter, and misusing words is being a poor steward of influence. Look to these daily questions to ensure you are stewarding your influence well:

1. Is my influence helping instructors be more effective?
2. Would I be motivated with my words?
3. How can my influence help shape my staff in a way that builds our school culture?

We need to use our influence to lift our teachers up so that they can do the same for our students. Consciously choose to lead staff down a path of success. Look to the positives — not negatives — of your team and students. Stewardship of influence is a powerful tool, so use it wisely and lead staff to obtainable goals.

Steward of Relationships

Relationships are a fundamental building block of leadership. As leaders, we are entrusted to build and maintain relationships with many stakeholders. First and foremost, we must build relationships with our students. Building professional, meaningful relationships with kids makes the entire educational process easier for all of us. Get to know your students, relate to them, and build them up.

We also must build relationships with our peers and faculty. We are all here for each other and our kids. Let's encourage one another and remember that our weaknesses are others' strengths. Learn from each other and grow.

The wise steward of relationships works with parents. Make contact and share successes so that when a failure happens, you have formed a relationship and can move forward together. Look to these questions as you steward leadership through relationships:

1. What are my weaknesses and how can my relationship strengthen those weaknesses?
2. Can my strengths help those around me to be better?
3. How can this relationship make me a better leader?

An educational leader can utilize stewardship of relationships in valuable ways. Leaders are most productive when building relationships. It is hard for anyone to be effective without actively pursuing growth in this area. We are not on an island all by ourselves, but we are here for each other and our students.

Burress suggests you may see a higher priority in one area over another, and asks that as the year begins, you reflect upon them and try to strengthen your stewardship as it relates to your students, staff, school, and community.

Burress's final comment is appropriate for teachers and students as it relates well to the chapter's definition of stewardship — Trust in yourself to be a wise steward and remember, the wise steward leaves the things they are entrusted with in better shape than they were given.

 One approach talking about stewardship, is to talk about it from the perspective of gifts and giving. Based on our definition that stewardship is taking care of what has been entrusted to you, discuss with students what they do with these "gifts"?

- You may ask students to talk about a time when they had something important, what it was and how did they take care of it because it was special?
- What motivated them to give it away?
- How did it make them feel to give it to someone else?
- Focus on the sharing aspect and that someone else was able to enjoy the gift. They took care of it and now someone else was able to enjoy and take care of it as well.

Earth Day — Smart Practices

Every year on April 22, Earth Day is celebrated marking the anniversary of the birth of the modern sentimental movement started in 1970.[8] In 1990, Earth Day became an International Day of Observance. Currently, over 192 countries observe Earth Day, and it is the largest secular observance in the world.[9]

Teaching students about Earth Day is vital to future generations understanding the importance of preserving the land on which we live. Each generation bears responsibility for what future generations will inherit. Education is the most powerful catalyst for change. Ideally, after learning about this day's importance, students will be inspired to act, whether that involves planting a tree, recycling at home, or picking up trash at a public park. We only get one Earth, so let's take care of it.[10]

Not taking care of the earth and the resources it provides is depriving future generations of its bounty. Big steps often start with little steps. If not already familiar with the very popular three R's related to the environment, this is a good place to start reinforcing the concept of stewardship. Introduce your students to reduce, reuse, and recycle.

- Make this a classroom activity by talking about packaging and waste.
- Discuss how items may be reused. Introduce the idea of repurposing items.
- Recycle what is appropriate as a classroom activity.
- Encourage parents and guardians to participate as they bring in classroom supplies and treats, and also by extending the three R's to the home.

Many classrooms already have projects such as growing plants indoor or outdoor gardens, or have incubators for animal propagation i.e. watching chicks hatch, etc. If not already doing this, planting seeds is an easy way to start. Most libraries have free seed libraries. Check to see if your local library provides free seeds you may "checkout" seeds to plant. It can be a fun and low-cost project.

Another practical activity is to discuss resources in your part of the world, which may be incorporated into science and geography studies.

Topic research may include identifying local, country and continental resources

The impact depletion would have on your area.

The need to respect and conserve these resources and ways to accomplish this, beginning in the classroom.

The peer-reviewed study "The role of climate change education on individual lifetime carbon emissions" investigated the long-term impact that an intensive 1-year university course had on individual carbon emissions by surveying students at least 5 years after having taken the course. Surveys and focus group interviews identify that course graduates have developed a strong personal connection to climate change solutions, and this is realized in their daily behaviors and through their professional careers. The analysis also demonstrates that if similar education programs were applied at scale, the potential reductions in carbon emissions would be of similar magnitude to other large-scale mitigation strategies, such as rooftop solar or electric vehicles.[11] Imagine if we begin the process of talking about good earth practices at an early age, what the children of today could do tomorrow.

The importance of stewardship has been presented to younger learners as the vigilant care of something valuable. This builds on previous lessons and activities to focus children's responsibility on safe online behavior, caring for possessions, and their environment. This foundational approach aims to cultivate future custodians of the Earth by teaching respect and care for the planet.

The concepts of respect, care, and environmental stewardship play an important role in the child's online environment as well.

The concept of digital stewardship and a young learner's role in this stewardship is an essential educational lesson for students. Educating children on environmental and digital stewardship can link and reinforce the idea of being environmentally aware of the planet around them and cyber-safe aware when in an online environment.

What Is Digital Stewardship?

The concept of digital stewardship and a student's role in this stewardship is an essential educational lesson for students. Educating children on environmental and digital stewardship can link and reinforce the idea of being environmentally aware of the planet around them and cyber-safe aware when in an online environment.

Digital stewardship refers to the responsible management and oversight of digital information and technology resources. In the realm of technology and cybersecurity, digital stewardship encompasses the practices and strategies employed to ensure that digital data is effectively managed, protected, and preserved over time. This includes the adoption of rigorous cybersecurity measures to safeguard information from unauthorized access, cyber threats, and data breaches. It also involves implementing data governance policies that dictate how data is collected, stored, accessed, and disposed of in a manner that respects privacy and compliance standards.

Furthermore, digital stewardship extends to the maintenance of the integrity and accessibility of digital assets throughout their lifecycle.

This requires a proactive approach to technology management, including regular software updates, data backup routines, and the deployment of robust encryption methods. Such practices ensure that not only is the data secure but also that it remains usable and meaningful for future applications.

Stewards of digital resources must also keep abreast of evolving technological landscapes, adapting their strategies to mitigate new risks and leverage emerging tools for better data management. As such, digital stewardship is a holistic endeavor, balancing the technical aspects of cybersecurity with the ethical and strategic management of digital information to uphold its value and utility over time.

Is It Digital Citizenship or Stewardship?

Digital citizenship and digital stewardship are related concepts but with distinct focuses. Leading a discussion with students, outlining the concepts of both while highlighting the differences may include addressing these key points...

Digital Citizenship (as discussed in Chapter 2)

- Digital citizenship refers to the responsible and ethical use of technology and the Internet.
- It encompasses concepts like online safety, cyberbullying prevention, digital literacy, respecting others' online privacy, understanding copyright laws, and being a positive digital presence.
- Key aspects include practicing good digital etiquette, being mindful of online behavior, protecting personal information, and understanding the consequences of online actions.

For third- and fourth-grade school children, the emphasis should be on fundamental digital citizenship concepts like:

a. Internet safety (such as not sharing personal information online.
b. Recognizing and avoiding online dangers.
c. Being respectful and kind in online interactions.
d. Understanding the basics of digital privacy.
e. Following rules for appropriate technology use.

Digital Stewardship

- Digital stewardship focuses on the responsible management and preservation of digital resources and the environment.
- It involves practices aimed at minimizing digital waste, reducing carbon footprints related to digital technologies, and promoting sustainability in digital practices.
- Key aspects include understanding the environmental impact of digital technologies, practicing responsible consumption and disposal of electronics, promoting energy-efficient use of devices, and advocating for sustainable digital practices.

For third and fourth-grade school children, the focus should be on simpler concepts like:

a. Understanding the importance of conserving energy by turning off devices when not in use.
b. Learning about electronic waste, and the importance of recycling electronics.
c. Being mindful of excessive screen time.

While digital citizenship focuses on responsible and ethical behavior in the digital realm, digital stewardship emphasizes responsible management and sustainability of digital resources.

For third- and fourth-grade school children, it's essential to introduce basic concepts of both, tailored to their developmental level and focusing on building a foundation of responsible digital behavior and environmental awareness.

Importance of Being Good Stewards in The Digital World

In discussing the role and importance of being a good steward in the digital world with students, addressing the concept simply, and directly and working with analogies that resonate with third- and fourth-grade students is a recommended approach.

A good start is to remind students that being a good steward in the digital world means taking care of their online spaces and the resources available to them just like they would take care of their toys and classrooms. It means being kind when they message or talk to others on the

Internet, not sharing personal information like their home address or secret safe word, and asking a trusted person for help if they see something online that makes them feel confused or uncomfortable.

Explaining Digital Stewardship to Third- and Fourth-Grade Students

One tactic to begin the discussion of being good digital stewards and practicing good cyber safety with third- and fourth-grade students may take this approach.

Digital stewardship is like being a superhero for computers and information on the Internet. It means taking good care of all the digital stuff you collect when you use your computer or are online like photos, games, and messages, so everything stays safe and works well.

Just like how you might protect a younger child, your sister, brother, or a friend on the playground, practicing being a good digital steward protects you and your online world.

The following are some cyber examples, which have been previously discussed, placed in the context of digital stewardship that students will recognize.

Creating Strong Passwords   Imagine your secret clubhouse where you keep your favorite toys. You don't want anyone sneaking in and taking them, right? So, you make a secret password that only you and your friends know. That's like creating a strong password for your online accounts to keep your information safe from hackers.

Updating Software   Think about when you get a new level in a video game that fixes old problems and adds cool new features. Updating software on computers and tablets is like that. It helps fix errors the programs have and makes sure everything is the best it can be to stop any cyberbullies from breaking in.

Backing Up Data   Have you ever made a cool LEGO build and then someone accidentally bumped into it, and it broke? That's no fun! Backing up data means making a copy of the important things you have on your computer. So, if the computer breaks or the information gets lost, you have a backup, just like keeping design instructions for your LEGOs safe.

Figure 4.3 Digital stewardship superhero.[12]

Being Kind Online

Just like in school, where you're taught to be nice to your friends and not say mean things, being a good digital steward means being kind and respectful online, too. You should always treat others how you want to be treated, even on the Internet.

By doing these things, you are being a Digital Stewardship Superhero! (see Figure 4.3). You are helping to protect your information and everyone else's, making the Internet a safer place to play and learn.

Digital Stewardship Framework

A framework, in a general sense, is a structured set of principles, guidelines, concepts, and standards that provide a foundation and systematic approach for understanding, designing, and implementing processes, projects, or systems.

Frameworks typically include a combination of best practices, methodologies, tools, and rules that aim to streamline and standardize the approach to achieving objectives, facilitating efficiency, consistency, and quality outcomes.

A "Digital Stewardship Framework" is a structured approach to managing the life cycle of digital information. It encompasses policies, strategies, and actions that ensure the accessibility, preservation, security, privacy, and usability of digital assets over time. The goal is to maintain the integrity and accessibility of digital information in a way that supports its long-term value, relevance, and availability for future use.

Good digital stewardship for students in the third and fourth grades could be conceptualized as a set of principles and practices that encourage responsible, respectful, and safe use of digital technology.

Educators may wish to approach a discussion of digital stewardship with students using the following framework.

Digital Stewardship Framework Principles for Third- and Fourth-Grade Students

A framework is a structured set of concepts, practices, and tools designed to provide guidance, standardization, and efficiency for developing software applications or solving complex problems in various fields such as engineering, business, or academia.

A framework for developing good digital stewardship refers to a structured approach or set of guidelines that organizations or individuals use to effectively manage and preserve digital assets over time. It encompasses policies, procedures, and best practices aimed at ensuring the long-term accessibility, integrity, and usability of digital information, including data, documents, and multimedia (see Figure 4.4).

The following presents Digital Stewardship Framework Principles that can be discussed with third- and fourth-grade students and form a basis of their overall cyber-safe practices.

Digital Stewardship Framework Principles

Respect Yourself and Others

- Be Kind Online — Like in the classroom or playground, always use kind words and actions.
- Respect Privacy — Do not share personal information about yourself or others.

Figure 4.4   Digital stewardship framework.[13]

Stay Safe

- Personal Information — Never give out personal information without a parent's, trusted adult's, or your teacher's permission. This includes your name, address, school, or phone number.
- Stranger Awareness — Just as you wouldn't talk to strangers in the park, don't talk to strangers online.

Protect Your Digital Footprint

- Think Before You Click — Understand that everything you do online can be seen by others and may stay online forever.
- Use Secure and Safe Websites — Learn how to identify websites that are safe for kids.

Use Resources Wisely

- Cite Your Sources — When you use information or pictures from the Internet, always give credit to the person who made them.
- Conserve Bandwidth — Use the Internet for school and learning purposes and understand that downloading large files can slow down the Internet for others.

Balanced Use of Technology

- Screen Time Limits — Know that it's important to balance screen time with other activities like playing outside, reading, and spending time with family.
- Healthy Habits — Take breaks from screen time to rest your eyes and move your body.

Be a Problem Solver

- Report Problems — If something online makes you feel uncomfortable or seems wrong, tell a trusted adult or your teacher right away.
- Learn to Troubleshoot — Learn some basic steps to solve common computer or Internet issues.

Learn and Create

- Educational Purpose — Use digital tools to learn new things and to work on school projects.
- Be Creative — Use technology to create art, and music, or to write stories.

Communicate Effectively

- Appropriate Communication — Use respectful language in emails, chats, and all online communication.
- Understand Emotions Online — Remember that without seeing someone's face, it can be hard to know their feelings. Use emojis wisely to help express your emotions.

Educators play a critical role in reinforcing these principles through regular discussion, modeling appropriate behavior, and guiding as students interact with technology. It's also important for educators (as well as parents, guardians, etc.) to stay informed about the latest in digital safety to guide their students effectively.

Developing a Stewardship Pledge

To foster awareness that emphasizes to students that this is a continual responsibility, teachers may work with students to develop a classroom stewardship pledge.

Discussions have centered around building character, making good choices, safe cyber habits, and the student's responsibility to act safely with possessions, with other people, and with the planet.

To create the classroom pledge, you may want to start by discussing respect and what it means. Respect takes many forms — to the student's belongings, to other people's belongings, to each other, to the community, and to the planet.

After discussing what respect means, a classroom pledge may be built on the word RESPECT.

Here's a classroom pledge that incorporates the concepts of character education, acceptable behavior, ethical behavior, practicing and maintaining safe cyber habits, and stewardship using the letters "R," "E," "S," "P," "E," "C," and "T."[14]

In the classroom, we pledge to:

R — "Reach out"

We promise to reach out to a teacher when we see something, such as project materials or snacks, being wasted in our classroom, to do what is good for our environment.

E — "Engage"

We will engage in kind behavior with my classmates, in person and online.

S — "Speak up"

We will speak up and let a teacher know when someone is being bullied or treated unkindly.

P — "Participate"

We commit to participate, in person and online, in kind and honest activities.

E — "Enjoy"

We will enjoy being good stewards of all we have and being a model to our school and community.

C — "Conduct"

We will conduct ourselves in a way that shows we are kind and respectful to each other and to our environment.

T — "Teach"

We promise to teach other classrooms and our friends about our pledge to increase kindness and good character in our school, at home, and in our community.

As the students begin to comprehend and grasp the classroom stewardship pledge, a personal student stewardship pledge can be developed. This pledge, developed at an early age, will lead to a sense of responsibility that will stay with the student into future years.

Here's a student pledge that incorporates the concepts of character education, acceptable behavior, ethical behavior, practicing and maintaining safe cyber habits, and stewardship using the letters "R," "E," "S," "P," "E," "C," and "T."[15]

R — "Respect"

"I pledge to show respect to everyone I meet, both online and offline."

E — "Ethical"

"I promise to be ethical and do what's right, even when no one is watching."

S — "Safe"

"I will keep myself safe online by not disclosing personal information about me or my family."

P — "Patient"

"I commit to being patient as I show friends and family what it means to take care of our community and the earth."

E — "Encourage"

"I will encourage others to take care of our environment and be a good steward of the Earth."

C — "Character"

"I promise to build good character by being honest, kind, and responsible."

T — "Trustworthy"

"I will be trustworthy and make good choices in everything I do."

This pledge helps younger children understand the importance of character education, acceptable and ethical behavior, practicing safe cyber habits, and stewardship while using simple words and concepts that are easy for kindergarten through second-grade students to comprehend and embrace.

Summary

The preceding section of this chapter focused on exploring the concept of cyber safety and responsible technology use through the lens of digital stewardship. Digital stewardship has been defined as the responsible management and oversight of digital information and technology resources.

The material presented in this section traces the historical evolution of the meaning of "stewardship" from managing estates in ancient times to its modern applications in areas like environmental protection, corporate governance, and now the digital realm.

In this chapter digital stewardship is distinguished from digital citizenship, with the former emphasizing responsible management and sustainability of digital resources, while the latter focuses more on ethical online behavior.

The importance of teaching digital stewardship to third- and fourth-grade students is emphasized by highlighting and using simple analogies they can relate to like protecting their digital "clubhouse" through strong passwords.

A proposed "Digital Stewardship Framework" that may be used by students in the third and fourth grades outlines key principles such as:

- Respecting self and others online
- Staying safe by not sharing personal information
- Protecting their digital footprint
- Using resources wisely by citing sources
- Balancing technology use with other activities
- Reporting problems and learning to troubleshoot
- Using technology for educational purposes and creativity
- Communicating effectively and appropriately online

The preceding material emphasizes the critical role educators play in reinforcing these digital stewardship principles through regular discussion, modeling appropriate behavior, and guiding students as they interact with technology.

Educators, parents, and guardians need to stay informed about the latest in digital safety to effectively guide their students.

LESSON PLANS

Grade 3

TOC Title: G3 Stewardship
  Lesson Title: Stewardship
  Grade Level: 3
  Duration: 50 minutes

Objective:

- The students will understand the importance of being responsible digital citizens.
- The students will demonstrate safe and responsible technology use through the concept of digital stewardship.

Materials Needed:

- Printed online safety scenarios (provided below).
- Poster-making materials (large paper or poster board, colored pencils, markers, or crayons).

Introduction (10 minutes):

a. Whole group discussion: ask students what technology devices they are familiar with, e.g., computers, tablets, and smartphones
b. Discuss the benefits of technology in our lives such as learning, entertainment, communication, etc.
c. Define the concept of digital stewardship as caring for the digital world by being responsible with digital information, staying safe online, and using digital resources responsibly.

Activity 1: Online Safety Scenarios (15 minutes):

a. Divide class into groups of 3 to 4, adjusting as needed based on class size.
b. Distribute online safety scenarios (included below).
c. Ask students to discuss in their groups how they would respond to each scenario.
d. Encourage them to consider the principles of digital stewardship and responsible technology use in their responses.
e. Have groups briefly share the results of their discussions with the class.

Activity 2: Digital Citizenship Poster (20 minutes):

    a. Break into groups, using the same groups from activity 1 or designate new groups of 3 to 4 students.

    b. Assign groups an aspect of digital citizenship discussed in class, e.g., being responsible with digital information, staying safe online, and the responsible use of digital resources.

    c. Distribute supplies for each group to make a poster.

    d. Groups create posters promoting their assigned aspect of digital citizenship.

    e. Teacher monitors groups offering guidance as needed. Refer to Chapter 4 Exploring Cyber Safety Through Digital Citizenship content as needed to help groups.

    f. Have groups tell the class about their poster, including how they incorporated the assigned aspects into the poster concept and design.

    g. Have groups hang their posters around the classroom.

Conclusion and Reflection (5 minutes):

    a. Recap the key points of the lesson. Emphasize the importance of being responsible digital citizens.

    b. Ask students to reflect on one thing they learned about digital stewardship and how they can apply it to their own technology use.

    c. Remind students that by being responsible digital citizens, they are helping to build a safer, more positive online environment for everyone.

Assessment:

    a. Observe students' participation in discussions and group activities. Assess their understanding of digital stewardship based on their contributions and reflections during the lesson.

Grade 4

TOC Title: G4 Stewardship
   Lesson Title: Stewardship
   Grade Level: 4
   Duration: 45 minutes

Objective:

- Students will understand the concept of digital stewardship.
- Students will learn simple strategies for practicing digital stewardship in their daily lives.

Suggested Materials:

- Whiteboard or chart paper.
- Markers.
- Printed list of digital stewardship scenarios (provided below).

Procedure

Introduction (5 minutes):

a. Full group discussion — start by asking students if they know what the term "stewardship" means.
b. Discuss briefly to ensure students understand stewardship as the responsible management and use of resources.
c. Explain that the term "digital stewardship" is about the responsible use of digital resources. Refer to Chapter 4 Exploring Cyber Safety Through Digital Citizenship content as needed.

Activity 1 Brainstorm (10 minutes):

a. Write the term "Digital Stewardship" on the board. Restate that it means being responsible with digital resources including computers, tablets, and the Internet.
b. Lead a discussion focused on why digital stewardship is important. Emphasize that taking care of digital devices and being mindful of how we use them can help make sure they last longer and are available for everyone to use. Add your own ideas about digital stewardship and managing them as a resource.

   c. Ask students to brainstorm ideas for practicing digital steward-
      ship. Write their ideas on the whiteboard. Ideas may include:
- Signing out and shutting down devices at the end of the school day.
- Being careful when using and transporting devices.
- Promptly reporting any technical issue.
- Using the Internet responsibly.
- Following school district rules for technology/Internet use.

Activity 2 Role-Playing (20 minutes):

   a. Divide the class into small groups (3–4 students per group)
     and assign each group a digital stewardship scenario (pro-
     vided below). Have them role-play how they would handle
     their situation responsibly.

   b. Discuss group responses as a class encouraging constructive
     advice, and appropriate questions after each group gives their
     response to their scenario. The teacher then offers positive/
     constructive feedback to groups.

Conclusion (10 minutes):

   a. Summarize key points discussed during the lesson.

   b. Reflection: Ask students to reflect on how they can apply what
     they've learned about digital stewardship at school and at home.

Assessment:

   a. Informally assess students' participation during discussions
     and activities.

Digital Stewardship Scenarios and Possible Answers:

Scenario: A student notices that a classroom device has been left on overnight.

Possible Answer: The student should inform their teacher or a school staff member immediately. Leaving devices on overnight wastes energy and reduces the lifespan of the device.

Scenario: Two students want to use the same computer at the same time.

Possible Answer: The students should take turns using the computer. They can practice patience and cooperation by agreeing on a fair amount of time for each person to use the computer.

Scenario: A student accidentally spills water on a school laptop.

Possible Answer: The student should notify a teacher or school staff member immediately. They should avoid attempting to turn on or use the laptop to prevent further damage.

Scenario: A student finishes using a school device and forgets to log out of their account.

Possible Answer: Whoever uses the device next should let the teacher know that someone else is still signed in. The teacher should remind the student why it is important to log out of a device when they are done using it. Logging out ensures privacy and security for their personal information.

Scenario: A student notices that a classroom printer is jammed.

Possible Answer: The student should inform their teacher or a school staff member. Attempting to fix the printer themselves may cause further damage.

Scenario: A student finishes using a shared classroom computer but leaves several unnecessary programs running.

Possible Answer: The student should close all unnecessary programs before leaving the computer. This ensures that the computer runs smoothly for the next user.

Scenario: A student encounters a website with inappropriate content while doing research.

Possible Answer: The student should close the website immediately and notify their teacher. It's important to follow school rules and guidelines for Internet usage and report any inappropriate content.

Scenario: A student wants to download a game onto a school computer.

Possible Answer: The student should ask their teacher or a school staff member for permission before downloading any software onto school computers. Unauthorized downloads can pose security risks and violate school policies.

Scenario: A student notices that a computer monitor is flickering.

Possible Answer: The student should inform their teacher or a school technician immediately.

Scenario: A student receives a phishing email asking for personal information.

Possible Answer: The student should not respond to, or forward the email. They should report it to a teacher or school staff member immediately. It is important to recognize and avoid phishing attempts to protect personal information and school data.

Notes

1 March 8, 2024, The Meaning of Stewardship - Historical Evolution, graphic developed and drawn by the authors.

2 (n.a.) "Responsibility for Your 10-Year-Old," Parenting Montana, Responsibility for Your 10-Year-Old.pdf, retrieved March 13, 2024.

3 Tucker, D., Bill, A., & Kathryn, H., 2023. "Student-driven Stewardship Projects Using environmental education to promote the use of science practices," National Science and Teaching Association, Student-driven Stewardship Projects | NSTA, www.nsta.org/science-and-children/science-and-children-septemberoctober-2020/student-driven-stewardship-projects, retrieved December 15, 2023.

4 (n.a.) "Raising Children to be Good Stewards of the Earth," Raising Children to be Stewards of the Earth | Lesley University, retrieved December 14, 2023.

5 Fletcher, C., October 5, 2023, "The Importance of Environmental Education for a Sustainable Future," Earth.org, The Importance of Environmental Education for a Sustainable Future | Earth.Org, retrieved December 15, 2023.

6 Rideout, V., et al, "The Common Sense Census: Media Use by Tweens and Teens, 2021," 8–18-census-integrated-report-final-web_0.pdf. commonsensemedia.org, Graphic drawn by authors from data presented in the above cited report, retrieved March 8, 2024.

7 Burress, D., May 15, 2020, "The Importance of Stewardship in Leadership," Graduate Programs for Educators, The Importance of Stewardship in Leadership - Graduate Programs for Educators, retrieved December 15, 2023.

8 https://www.earthday.org/, retrieved November 3, 2023.

9 Earth Day 2023: 5 Earth Day Facts to Know About and How To Get Involved | Earth.Org, retrieved December 15, 2023.

10 (n.a.) "Teaching About Earth Day," Educators4SC, Teaching About Earth Day - Educators 4SC, retrieved December 19, 2023.

11 Cordone, Eugene C., Centeno, D., & Todd, A. M., February 4, 2020, "The role of climate change education on individual lifetime carbon emissions." *PLOS ONE*, retrieved March 1, 2024.

12 Digital Stewardship Superhero, Shutterstock Stock Photo ID 743404276 used under CRC license with Shutterstock

13 March 13, 2024, Digital Stewardship Framework, graphic developed and drawn by the authors.

14 Marcella, A., "Create a student pledge summarizing the concepts of character education, acceptable and ethical behavior, practicing and maintaining safe cyber habits, and stewardship," ChatGPT-4, developed by OpenAI, retrieved January 4, 2023.

15 Marcella, A., "Create a student pledge summarizing the concepts of character education, acceptable and ethical behavior, practicing and maintaining safe cyber habits, and stewardship," ChatGPT-4, developed by OpenAI, retrieved January 4, 2023.

Appendix A: Teacher's Resource List

The following materials represent a non-exhaustive list of supplemental teaching resources that the educator may find both of interest and use in developing class instructional materials to further the cyber-safe education of young learners. The resources are presented alphabetically, and the URL included with each resource is the most current as of May 2024.

Blogs

Sleepovers in the Digital Age, Haley Zapal, February 25, 2019, The Bark Blog. A deep dive into the pros and cons of taking phones away at sleepovers and provide a path that may satisfy both camps.

Children's Illustrated Books

"Bullies Never Win" by Margery Cuyler and Arthur Howard — Standing up for oneself is the theme of this book. {ISBN 13: 978-0689861871}

"Kindness is Cooler, Mrs. Ruler" by Margery Cuyler and Sachiko Yoshikawa is another illustrated book that is great for starting conversations about kindness, a key element of character education. [ISBN 13: 978-0689873447}

"Kidpower Children's Safety Comics Color Edition: Use Your Power to Stay Safe!" Have fun while teaching kids to be safe from bullying, abuse, and violence. The books have updated social stories and skills, including how to help kids stay safe online. [www.kidpower.org/books/kidpower-childrens-safety-comics-color-edition]

"Stand in My Shoes: Kids Learning about Empathy" by Bob Sornson — This book, while not a comic, uses engaging illustrations and storytelling to teach kids about empathy. [ISBN 13: 978-0578807942]

Cyber-Safety Research

Finkelhor, D., Walsh, K., Jones, L., Mitchell, K., & Collier, A., 2021, "Youth internet safety education: Aligning programs with the evidence base," *Trauma, Violence, & Abuse*, Volume 22, Issue 5, 1233–1247. https://doi.org/10.1177/1524838020916257. [This research reviews the literature on online harms to children and discusses how to develop effective online safety messages.]

Hartikainen, H., Iivari, N., Kinnula, M., December 2019, "Children's design recommendations for online safety education," *International Journal of Child-Computer Interaction*, Volume 22, 100146, ISSN 2212-8689, https://doi.org/10.1016/j.ijcci.2019.100146, www.sciencedirect.com/science/article/pii/S2212868917300764. [Focuses on involving children in the design of online safety education, ensuring that the content is relatable and understandable for them.]

Moynihan, M., November 28, 2014, *Using Web 2.0 Tools to Teach Online Safety Education in the Intermediate Grades*, Vancouver Island University, https://viuspace.viu.ca/bitstream/handle/10613/2328/MoynihanOLTD.pdf?sequence=3. [Demonstrates how Web 2.0 tools can be effectively used in teaching online safety to young students.]

Ondrušková, D., Pospíšil, R., 2023, "The good practices for implementation of cyber security education for school children," *Contemporary Educational Technology*, Volume 15, Issue 3, Article No. ep435, ISSN: 1309-517X (Online), https://doi.org/10.30935/cedtech/13253. [Outlines best practices for implementing cybersecurity education in schools, emphasizing age-appropriate methods.]

Wishart, J., 2004, "Internet safety in emerging educational contexts," *Computers & Education*, Volume 43, 193–204. https://doi.org/10.1016/j.compedu.2003.12.013, www.researchgate.net/publication/220140831_Internet_safety_in_emerging_educational_contexts/citation/download.

[This paper reports on the consequent audit of Internet Safety practices in over 500 schools from 27 Local Education Authorities (LEAs) across England, commissioned by Becta and conducted during the summer term 2002.]

Digital Resources

BrainPOP Jr. Internet Safety Video: A resource for teaching children about interacting with strangers online and protecting private information. Includes activities such as Word Play and Write About It, along with quizzes for assessment. [https://jr.brainpop.com/artsandtechnology/technology/internetsafety]

Georgia Public Broadcasting (GPB): GPB Education is Georgia's digital media content provider for the classroom offering locally produced, Georgia-specific content and digital streaming services across all subject areas to teachers and students. GPB's goal is to remain at the forefront of the digital learning movement by creating, curating, and distributing quality educational programs and services through a state-of-the-art production facility, cutting-edge digital media division, and a partnership with PBS LearningMedia. GPB Education also provides training for Georgia's educators and educational institutions. [www.gpb.org/education/learn/school-stories/science-in-action]

Mentoring Youth: Provides resources and tools for mentoring youth. [https://youth.gov/youth-topics/mentoring#:~:text=Mentoring%20provides%20youth%20with%20mentors,DuBois%20and%20Karcher%2C%202005)]

National Alliance on Mental Illness (NAMI): In addition to NAMI's support of mental illness, they provide resources for bullying, a teen and youth call center, stress management, and much more. [www.nami.org/Home]

Neuroscience in Education — Center for Educational Neuroscience: A component of Mind Brain and Education is an emerging scientific field that brings together researchers in cognitive neuroscience, developmental cognitive neuroscience, educational psychology, educational technology, education theory, and other related disciplines to explore the interactions between biological processes and education. [www.educationalneuroscience.org.uk/about-us/what-is-educational-neuroscience]

Instructional Resources

Common Sense Education — 23 Great Lesson Plans for Internet Safety: Provides lesson plans to help students build critical-thinking skills and navigate online dilemmas. [www.commonsense.org/education/articles/23-great-lesson-plans-for-internet-safety]

Common Sense Education Safety in My Online Neighborhood: offers lesson plans for digital citizenship that address timely topics such as cyberbullying, cybersecurity, identity verification, physical security, and online safety. Their curriculum is research-backed and designed to help schools navigate these issues with their students. Lesson plans culminate in interactive activities like the Internet Traffic Light game, where students assess whether the presented online content is appropriate or not. [www.commonsense.org]

Edutopia: provides a lesson plan for teaching Internet safety to students as young as kindergarten. The plan includes discussions about strangers online and protecting private information. Additional activities such as watching an Internet safety video, quizzes, and creating safety posters are also outlined. [www.edutopia.org]

EVERFI: The Compassion Project: A free empathy curriculum for 2nd–5th graders, designed to facilitate lessons around social and emotional skills. [https://everfi.com/k-12/character-education]

GoodCharacter.com — Character Education and Social-Emotional (SEL) Learning Resources: Provides teaching guides for K-12 on character education and social-emotional learning, covering topics like mindfulness, conflict resolution, and bullying.

- Bullying [www.livewiremedia.com/product-category/bullying]

- Citizenship Guides [www.livewiremedia.com/product-category/citizenship]
- Conflict Resolution [www.livewiremedia.com/product-category/conflict-resolution-2]
- Fairness [www.livewiremedia.com/product-category/fairness]
- Good Character Teaching Guides [www.livewiremedia.com/product-category/character-education] and [www.livewiremedia.com/product-category/character-counts]
- Manners and Politeness [www.livewiremedia.com/product-category/manners-politeness]
- Mindfulness [www.livewiremedia.com/product-category/mindfulness]
- Respect [www.livewiremedia.com/product-category/respect]
- Social Media & Internet Safety [www.livewiremedia.com/product-category/social-media-internet-safety-2]

Safer, Smarter Kids Curriculum: A comprehensive program that covers various safety topics, including cyber safety, body boundaries, and recognizing unsafe situations. [https://safersmarterkids.org]

TeacherVision Safety Printables and Activities: Offers a wide array of lesson plans, printables, and resources, including content for teaching Internet safety. [www.teachervision.com/subjects/health-safety/safety]

UNODC Primary Lesson Plan for Staying Safe Online: While this resource is more general, it does include exercises and additional teaching tools that can be adapted for younger students. [www.unodc.org/e4j/en/primary/e4j-tools-and-materials/lesson-plan_staying-safe-online.html]

Interactive Assessment Exercises

Cyber Awareness & Safety Education (CASE): This curriculum offers a series of interactive classes where students actively learn through videos, case studies, and group discussions. The end-of-course test measures mastery of the material. This program includes pretests and post-tests to assess student understanding of Internet safety, along with active learning techniques to personalize the topics covered. [www.cybersafe-tyconsulting.com]

Interactive Games and Classroom Exercises

CISA Cybersecurity Awareness Program Student Resources: Offers downloadable activity sheets, books, and other resources for young children on cyber safety. It also includes games from the FBI's Safe Online Surfing program and tips from the National Cyber Security Alliance.

- Savvy Cyber Kids [https://savvycyberkids.org]
- Safe Online Surfing by FBI [https://sos.fbi.gov/en]
- NetSmartz Kids [www.netsmartzkids.org]

Digital Wellbeing Packet and Classroom Activities from Be Internet Awesome: is a multifaceted program that includes both a game called Interland to engage students in learning about online safety and a curriculum for educators to teach kids how to be safe and responsible online. They also offer a pledge for families to promote online safety at home.

The site provides comprehensive guides and printable activities to support educators in introducing digital well-being to students. Lesson plans for five topics with activities and worksheets designed to complement the Interland game, focusing on being smart, alert, strong, kind, and brave online. Provides interactive, game-based activities that assess students' understanding of Internet safety in a fun and engaging way. [https://beinternetawesome.withgoogle.com/en_us/educators]

Safety Land by AT&T: An interactive game for elementary students that teaches Internet safety in a fun and engaging way. [https://freetech4teach.teachermade.com/2011/04/at-sa fety-land-cyber-safety-game-for]

Publications

15-Minute Focus: Digital Citizenship: Supporting Youth Navigating Technology in a Rapidly Changing World

Raychelle Cassada Lohmann and Angie C. Smith
National Center for Youth Issues
May 4, 2023
ISBN 13: 978-1953945778

An Educator's Guide to AI in the Classroom: The Transformative Power of AI in Education, How to Use AI in School, K-12 Classroom Lesson Plans, and Answers to Common AI Questions
Abigail McKeag
ViaNova Productions, LLC
July 24, 2023
ISBN 13: 979-8988586296

Developing Digital Detectives: Essential Lessons for Discerning Fact from Fiction in the "Fake News" Era
Jennifer LaGarde, Darren Hudgins
International Society for Technology in Education
September 27, 2021
ISBN 13: 978-1564849052

Digital Citizenship Handbook for School Leaders: Fostering Positive Interactions Online
Mike Ribble and Marty Park,
International Society for Technology in Education
June 26, 2019
ISBN 13: 978-1564847829

Ethics in a Digital World: Guiding Students Through Society's Biggest Questions
Kristen Mattson
International Society for Technology in Education
April 19, 2021
ISBN 13: 978-1564849014

TED-Ed Video Resources

Note: TED-Ed is TED Talk's [www.ted.com/talks] youth and education initiative.

"3 Questions to Ask Yourself About Everything You Do" by Stacey Abrams: "How you respond to setbacks is what defines your character," a talk that could be adapted to discuss character and self-reflection, important components of character education. [https://ed.ted.com/search?qs=%223+questions+to+ask+yourself+about+everything+you+do%22+]

"Grit: The Power of Passion and Perseverance" by Angela Lee Duckworth: this talk may be adapted to be used to discuss the character trait of perseverance. [https://ed.ted.com/search?qs=Grit%3A+The+Power+of+passion+and+perseverance]

"How to Build Your Confidence — and Spark It in Others" by Brittany Packnett: This talk can help in discussions with students about self-confidence as a part of one's character. [https://www.youtube.com/watch?v=b5ZESpOAolU&t=4s]

YouTube Video Resources

"Being Respectful Video for Kids"|Character Education – by Jessica Diaz: This video can help teach kids about being respectful, an essential aspect of character education. [https://www.youtube.com/watch?v=KxnxObAyfSA]

"Consequences for Kids"|Character Education – by Jessica Diaz: This video is designed to help children understand what consequences are and why they matter, reinforcing the concept of accountability. [www.youtube.com/watch?v=LLZZYf_mlOA]

"Cyber Security for Kids" – by Neel Nation: This video is about being safe on the Internet. The video provides five tips for kids to be safe on the Internet. Being cyber-safe is very important as the kids are always online these days. This video talks about the importance of passwords, protecting personal information, limiting screen time, etc. [www.youtube.com/watch?v=nVEyG3C-Mqw]

"Cyberbullying – How to Avoid Cyber Abuse" – by Smile and Learn: This video showcases a series of situations related to how students can stop cyberbullying. Also discussed is the concept of being a good friend, not a bystander, and respecting privacy, and that actions can make a difference. [www.youtube.com/watch?v=dMdKmHjpgFk]

"Cybersecurity Training for Kids" – by Malware Bytes: A fun video just for the kids, where they can learn all about Dr. Evil and her Internet schemes. [www.youtube.com/watch?v=XiU72Vzs5Is]

"Internet Safety for Kids K-3" – by Indiana University of Pennsylvania: This video is aimed at younger children and provides guidance on staying safe on the Internet. [www.youtube.com/watch?v=89eCHtFs0XM]

"Internet Safety Tips for Kids" – by MoneyMoments (MidFirst Bank): This short video covers safety tips for your kids using the Internet. [www.youtube.com/watch?v=qtJNRxMRuPE]

"Online Privacy for Kids - Internet Safety and Security for Kids" – by Smile and Learn: This video discusses the importance of online privacy and safety for children, which is crucial for their well-being in the digital age. [www.youtube.com/watch?v=yiKeLOKcltw]

"Online Safety Staying Safe Online" – by Born in the USA: This resource features in Discovery Education Espresso's Online safety module. Learn how to stay safe and act responsibly when using the Internet in school and at home. [www.youtube.com/watch?v=PtfEnh0gbbU]

"Private and Personal Information" – by Common Sense Education: It is natural for students to enjoy sharing and connecting with others. However, sharing information online can sometimes come with risks. This video addresses these risks. [www.youtube.com/watch?v=MjPpG2e71Ec]

"Responsible Use of Technology for Kids" – by Smile and Learn: This video discusses how to make responsible use of technology, the Internet, and social media. In this compilation of videos about the responsible use of technology, the little ones will discover how to use their first phone responsibly, how to avoid cyberbullying, how to know what fake news is and how to spot it, and also how to protect their online privacy. [www.youtube.com/watch?v=JkkTN0pQ_Ug]

"What Is NETIQUETTE?" Internet Behavior Rules for Kids, Episode 1 – by Smile and Learn: by Smile and Learn.

This presentation discusses what netiquette is and how it can help children behave and interact correctly on the Internet. In this first series of two videos, kids will learn a basic set of rules to be polite and respectful on social networks and on the Internet in general. [www.youtube.com/watch?v=kZOfLN4YqhY]

"What is NETIQUETTE?" Internet Behavior Rules for Kids, Episode 2 – by Smile and Learn: In this second video of the series, kids will learn how to interact with other people on the Internet and see some tips on how not to share personal information or trust strangers. [www.youtube.com/watch?v=zhIm-CDJBpc]

Websites

Department of Justice (StopBullying.gov). Although no federal law directly addresses bullying, in some cases, bullying overlaps with discriminatory harassment when it is based on race, national origin, color, sex (including sexual orientation and gender identity), age, disability, or religion.

LEGO — Build and Talk — Sustainability (LEGO.com). Covering six different topics, the activities are designed to help children navigate the online world safely. You'll also find handy discussion starters to get the conversation going as you Build & Talk together.

Office of Juvenile Justice and Delinquency Prevention (OJJDP) School-Based Bullying Prevention I-Guide (ojp.gov). This Model Program I-Guide is focused on the problem of bullying in school and is intended for those interested in creating safe school environments by implementing school-based bullying prevention programs.

Smart Social [SmartSocial.com] Digital Citizenship programs. Resources to help parents & educators keep students safe on social media. Membership required.

Appendix B: Recommended Readings

The following titles represent a non-exhaustive list of recommended readings that the educator may find both of interest and use in furthering the cyber-safe education of their students. These reading resources are presented alphabetically by title, by section, and the ISBN 13 for each title (where appropriate) is the most current as of May 2024.

Books

"**Anti-Bullying Book for Girls: Practical Tools to Manage Bullying and Build Confidence" by Jessica Wood**: This standout among bullying books for kids will help girls find their voice and put a stop to bullying, whether it's happening to them or their friends. [ISBN 13: 978-1638079118]

"**CookBully B.E.A.N.S.: A Picture Book to Help Kids Stand Up Against Bullying" by Julia Cook**: *Bully B.E.A.N.S* helps children identify bullying and offers clear and impactful action strategies for both targets and bystanders. [ISBN 13: 978-1937870591]

"**Developing Mentoring and Coaching Relationships in Early Care and Education: A Reflective Approach" (Practical Resources in ECE) by Marilyn Chu**: is the ideal resource for anyone charged with guiding teachers as they encounter real-world challenges in today's early childhood programs. [ISBN 13: 978-0132658232]

"Disconnected: How to Protect Your Kids from the Harmful Effects of Device Dependency" by Thomas Kersting: [ISBN 13: 978-1540900302]

"Eesha and the Mud Monster Mystery" by Wendy Goucher: Join Eesha's cyber adventure to protect children from texting bullying, (ages 7-11) [ISBN 13: 978-1915641175]

"Growth Mindset Workbook for Kids: 55 Fun Activities to Think Creatively, Solve Problems, and Love Learning" (Health and Wellness Workbooks for Kids) by Peyton Curley: A growth mindset means practicing flexible thinking and looking at challenges as opportunities to learn and grow! Discover the power of a growth mindset with hands-on activities for ages 8–12. [ISBN 13: 978-1646117031].

"Hold On to Your Kids: Why Parents Need to Matter More Than Peers" by Gordon Neufeld, and Gabor Mate MD: [ISBN 13: 978-0375760280]

"How to Raise Healthy Gamers: End Power Struggles, Break Bad Screen Habits, and Transform Your Relationship with Your Kids" by Alok Kanojia MD MPH: [ISBN 13: 978-0593582046]

"Life Skills for Kids: How to Cook, Clean, Make Friends, Handle Emergencies, Set Goals, Make Good Decisions, and Everything in Between" by Karen Harris: (ages 8–10) [ISBN 13: 978-1951806446]

"Make Your Brain Work" by Amy Brann: While this book Is not for the 3rd and 4th grade student, it is a helpful tool for teachers and leaders to understand the latest insights from neuroscience about how our mind works. [ISBN 13: 978-1789660494]

"Critical Thinking, Logic & Problem Solving: The Complete Guide to Superior Thinking, Systematic Problem Solving, Making Outstanding Decisions, and Uncover Logical Fallacies Like a Pro" by Neurowaves Labs: [ISBN 13: 979-8866530397]

"Oh No …Hacked Again! A Story About Online Safety" by Zinet Kamal: While enjoying the adventures of online games, Elham struggles with making the safest decisions. [ISBN 13: 978-1737775911]

"Online Safety for Children and Teens on the Autism Spectrum: A Parent's and Care's Guide" by Nicola Lonie: While this is more of a guide for parents and caregivers, it offers valuable insights and strategies that can be translated into simpler concepts for young children. [ISBN 13: 978-1849054425]

"Once Upon a Digital Story" by Amanda Hovious: This book is designed to help children learn to tell their own stories safely and creatively online. [ISBN 13: 978-1516553341]

"Polar Bear, Why Is Your World Melting?" by Robert E. Wells: [ISBN 13: 978-0807565995]

"Security Awareness Design in the New Normal Age" by Wendy Goucher: People working in our cyber world have access to a wide range of information including sensitive personal or corporate information which increases the risk of it. This book will primarily consider the knowledge about secure practice is not only understood and remembered but also reliably put into practice — even when a person is working alone. [Kindle ASIN - B0B1W1Z9LB]

"The Anxious Generation: How the Great Rewiring of Childhood Is Causing an Epidemic of Mental Illness" by Jonathan Haidt: In The Anxious Generation, the author explains the major causes of the international epidemic of mental illness that hit adolescents in the early 2010s and offers a path forward for parents, teachers, friends, and relatives who want to help improve the mental health of children and adolescents. [ISBN 13: 978-0593655030]

"The Buzz About Social Media: A Cyber Safety Workbook and Discussion Guide for Pre-Teens, Ages 8–12" by Shawn Marie Edgington and Emily Scheinberg: The Buzz About Social Media was created to guide you through important discussions and activities related to cyber safety and proper online behavior. This workbook is filled with suggestions, real-life situations, and activities to help you discuss the best practices for cyber safety with your children and students. [ISBN 13: 978-1466495210]

"The Digital Citizenship Handbook for School Leaders: Fostering Positive Interactions Online" by Mike Ribble and Marty Park: Aimed at educators, contains valuable resources that can be adapted into lesson plans for young students. [ISBN 13: 978-1564847829]

"The No More Bullying Book for Kids: Become Strong, Happy, and Bully-Proof" by Vanessa Green Allen MEd NBCT: This engaging book gives kids the information they need to identify bullying, followed by strategies for dealing with specific situations when they or someone they know is being bullied. [ISBN 13: 978-1641520713]

"Young Native Activist: Growing Up in Native American Rights Movements (Young Native Boy Series Book 1)" by Aslan Tudor: Learn about activism and how young children have a voice through young Aslan's true story. [ISBN 13: 978-1074524746].

"Young Water Protectors: A Story About Standing Rock (Young Native Boy Series Book 2)" by Aslan Tudor: Read about the author's inspiring experiences in the Oceti Sakowin Camp at Standing Rock. [ISBN 13: 978-1723305689].

Research (by Lead Author)

Anitha, F. S., Narasimhan, U., Janakiraman, A., Janakarajan, N., and Tamilselvan, P. (2021). Association of digital media exposure and addiction with child development and behavior: a cross-sectional study. *Indian Journal of Psychiatry*, Volume 30, 265–271. https://doi.org/10.4103/ipj.ipj_157_20

Borajy, S., Albkhari, D., Turkistani, H., Altuwairiqi, R., Aboalshamat, K., Altaib, T., et al. (2019). Relationship of electronic device usage with obesity and speech delay in children. *Family Medicine & Primary Care Review*, Volume 21, 93–97. https://doi.org/10.5114/fmpcr.2019.84542

Domoff, S. E., Borgen, A. L., and Radesky, J. S. (2020). Interactional theory of childhood problematic media use. *Human Behavior and Emerging Technologies*, Volume 2, 343–353. https://doi.org/10.1002/hbe2.217

Dong, C., and Mertala, P. (2021). Preservice teachers' beliefs about young children's technology use at home. *Teaching and Teacher Education*, Volume 102, 103325. https://doi.org/10.1016/j.tate.2021.103325

Hutton, J. S., Huang, G., Sahay, R. D., DeWitt, T., and Ittenbach, R. F. (2020). A novel, composite measure of screen-based media use in young children (screen Q) and associations with parenting practices and cognitive abilities. *Pediatric Research*, Volume 87, 1211–1218. https://doi.org/10.1038/s41390-020-0765-1

Moon, J. H., Cho, S. Y., Lim, S. M., Roh, J. H., Koh, M. S., Kim, Y. J., et al. (2019). Smart device usage in early childhood is differentially associated with fine motor and language development. *Acta Paediatrica*, Volume 108, 903–910. https://doi.org/10.1111/apa.14623

The World Health Organization (WHO). (2019). "WHO Guidelines on Physical Activity, Sedentary Behaviour and Sleep for Children under 5 Years of Age," Very limited daily screen time is recommended for children under five. Available at: www.aoa.org/news/clinical-eye-care/public-health/screen-time-for-children-under-5?sso=y, [978-92-4-155053-6].

Appendix C: Ideas for Certificates or Badges

The following is a non-exhaustive list of popular platforms for creating digital badges. The sites are presented alphabetically, and the URL included with each site is the most current as of May 2024.

- Google Slides: https://docs.google.com/presentation
- Open Badges by Mozilla: https://openbadges.org
- Credly: www.credly.com
- Badgr: https://badgr.com
- Accredible: www.accredible.com
- Canva: https://www.canva.com
- Adobe Spark: https://spark.adobe.com

These platforms can help you create, issue, and manage digital badges or certificates to recognize and reward students for demonstrating good character skills, acceptable behavior, and safe cyber habits.

Certificates/Badges/Awards/Recognition Ideas

Creating certificates and badges for a *character education program* is a great way to recognize and encourage positive traits in third- and fourth-grade students.

Each of these awards can be designed with engaging graphics that reflect the quality they're celebrating, such as a bright star for the Perseverance Star or a gentle heart for the Inclusivity Heart. These badges and certificates can serve as tangible reminders for students of the values they are being encouraged to internalize and exhibit in their daily lives.

Character Education

Here is a list of topics and ideas that could be used for *Character Education* Certificates/Badges/Awards/Individual or Team Recognition Ideas:

- Honesty Award: For students who consistently display truthfulness in their words and actions.
- Kindness Medal: For those who are helpful and kind to others.
- Responsibility Ribbon: For students who take responsibility for their actions and duties.
- Respect Badge: Awarded to students who show respect to peers, teachers, and school property.
- Teamwork Trophy: For students who work well in a group and contribute to team efforts.
- Perseverance Star: For those who show determination and keep trying, even when things are difficult.
- Inclusivity Heart: For students who include others and embrace diversity.
- Compassion Certificate: For showing empathy and caring for the feelings of others.
- Courage Crest: For students who stand up for what is right or try new things despite fears.
- Integrity Shield: For making honest choices, even when no one is watching.
- Citizenship Crown: For participating in community-related activities and showing civic pride.
- Gratitude Badge: For consistently saying thank you and showing appreciation.
- Leadership Laurel: For students who take the lead in positive ways within the classroom or school activities.

- Fairness Flag: For those who play by the rules and ensure others are treated fairly.
- Environmental Steward Badge: For students actively engaged in recycling and taking care of the school environment.
- Conflict Resolution Medal: For students who help resolve disagreements peacefully.
- Helping Hand Award: For those who are always ready to assist others in need.
- Creative Thinker Badge: For showing originality and creativity in work and play.
- Mindfulness Medal: For students practicing mindfulness and self-regulation.
- Problem Solver Star: For displaying critical thinking and problem-solving skills.
- Positive Attitude Pin: For students who consistently approach challenges with positivity.
- Active Listener Ears: For students who listen carefully to others without interrupting.
- Punctuality Pocket Watch: For students who consistently arrive on time and meet deadlines.
- Self-Discipline Sash: For demonstrating control in managing, one's behavior and work.
- Patience Pendant: For waiting calmly and dealing with frustrations gracefully.

Acceptable and Ethical Behavior

Here is a list of topics and ideas that could be used for *Acceptable and Ethical Behavior* Certificates/Badges/Awards/Individual or Team Recognition:

These ideas focus on the positive behaviors you want to reinforce and recognize in students, and they can be visually represented in a way that is both engaging and meaningful for students.

Each badge or certificate can feature symbols or images that are associated with the behavior it rewards, like a handshake for the Trustworthiness Trophy or an elephant for the Empathy Elephant Medal, making them memorable and significant.

1. Honor Roll Badge: For consistently displaying honesty and integrity.
2. Golden Rule Award: For treating others as one would like to be treated.
3. Digital Citizen Badge: For responsible and respectful online behavior.
4. Eco-Warrior Award: For promoting and practicing environmental ethics.
5. Good Sports Medal: For showing sportsmanship, fair play, and grace in winning or losing.
6. Community Helper Badge: For engaging in actions that benefit the community.
7. Trustworthiness Trophy: For being reliable and trustworthy in different situations.
8. Peacekeeper Award: For mediating conflicts and helping maintain a peaceful school environment.
9. Animal Friend Badge: For showing care and ethical treatment toward animals.
10. Privacy Protector Medal: For respecting the privacy of others and keeping personal information safe.
11. Accountability Award: For taking responsibility for one's actions and understanding their consequences.
12. Cyber Safety Shield: For learning and following the rules of Internet safety and cybersecurity.
13. Responsible Researcher Badge: For citing sources and avoiding plagiarism.
14. Bullying Prevention Star: For actively standing against bullying and supporting affected peers.
15. Ethical Thinker Award: For making decisions based on what's right and fair.
16. Respect for Resources Ribbon: For using materials wisely and not wasting supplies.
17. Informed Citizen Medal: For staying informed on school rules and current events in an age-appropriate manner.
18. Caring Community Member Certificate: For demonstrating care for others in the school and local community.
19. Law of the Land Badge: For understanding and respecting classroom and school rules.

20. Sincerity Sunflower Award: For genuine and sincere interactions with others.
21. Patience Pebble Badge: For showing patience in waiting and dealing with challenges.
22. Empathy Elephant Medal: For showing understanding and empathy toward peers' feelings.
23. Justice and Fairness Badge: For ensuring fairness in play and classroom activities.
24. Consent Champion Ribbon: For understanding and respecting personal boundaries.
25. Anti-Cheating Champion Badge: For upholding academic honesty and encouraging others to do the same.

Safe Cyber Habits

The following is a list of topics and ideas that could be used for *Safe Cyber Habits* Certificates/Badges/Awards/Individual or Team Recognition Ideas:

Each badge or certificate can be designed with a visual element that represents its specific area of cyber safety. For example, a shield for the Privacy Protector Certificate or a magnifying glass for the Cyber Sleuth Award. These visuals can help students associate the images with the behavior or skill being recognized, reinforcing the lessons learned in the classroom.

1. Cyber Safety Scout Badge: For demonstrating knowledge of safe Internet browsing practices.
2. Privacy Protector Certificate: For understanding the importance of keeping personal information private.
3. Password Guardian Badge: For creating strong passwords and keeping them secret.
4. Cyber Bullying Defender Award: For recognizing cyberbullying and knowing how to respond.
5. Digital Footprint Award: For understanding that online actions are permanent and can be tracked.
6. Smart Surfer Badge: For showing discernment in evaluating the credibility of online information.
7. Netiquette Star: For consistently demonstrating polite and respectful online communication.

8. Anti-Phishing Medal: For identifying and avoiding phishing scams and suspicious links.

9. Responsible Posting Ribbon: For thinking before posting and understanding the impact of online words.

10. Screen Time Manager Award: For practicing balanced and healthy screen time habits.

11. Online Community Contributor Badge: For positively contributing to online communities and forums.

12. Security Champion Certificate: For staying updated with the latest in online security practices.

13. Tech Helper Badge: For assisting others in learning safe cyber habits.

14. Software Savvy Award: For understanding the importance of updates and legitimate software.

15. Creative Commons Advocate Badge: For respecting copyrights and using free-to-use resources properly.

16. Cyber Ethics Emblem: For discussing ethical dilemmas and making ethical decisions online.

17. Digital Literacy Laureate: For showing proficiency in using the Internet and digital tools responsibly.

18. E-Resilience Ribbon: For coping with online challenges and knowing when to seek help.

19. App Analyst Badge: For evaluating apps for safety and appropriateness before downloading.

20. Information Literacy Award: For distinguishing between facts and opinions online.

21. Online Safety Ambassador Badge: For leading by example and promoting safe cyber habits.

22. Gaming Guru Certificate: For practicing safety and fairness in online games and virtual environments.

23. Digital Detox Challenge Medal: For participating in activities that involve taking breaks from digital devices.

24. Virtual Reality Virtuoso Badge: For understanding virtual space and its potential risks.

25. Cyber Sleuth Award: For learning to investigate and understand the digital world safely.

Glossary of Terms & Definitions

Accessibility

Refers to the design of apps, devices, materials, and environments that support and enable access to content and educational activities for all learners. In addition to enabling students with disabilities to use content and participate in activities, the concepts also apply to accommodate the individual learning needs of students, such as English language learners, students in rural communities, or from economically disadvantaged homes. Technology can support accessibility through embedded assistance, for example, text-to-speech, audio and digital text formats of instructional materials, programs that differentiate instruction, adaptive testing, built-in accommodations, and assistive technology.

Artificial Intelligence

A system that exhibits reasoning and performs some sort of automated decision-making without the interference of a human.

Authentic Learning Experiences

Experiences that place learners in the context of real-world experiences and challenges.

Behavior

A person's external reaction to their environment. All behavior, such as yelling, crying, running, or throwing something, can be observed and measured.

Belittlement

The use of online forums, apps, or doctored images to try to insult and spread fake rumors, gossip, or untruths about a person.

Bioregion

An area whose limits are naturally defined by topographic and biological features (such as mountain ranges and ecosystems) and not political or religious boundaries.

Blended Learning

In a blended learning environment, learning occurs online and in person augmenting and supporting teacher practice. Blended learning often allows students to have some control over the time, place, path, or pace of learning. In many blended learning models, students spend some of their face-to-face time with the teacher in a large group, some face-to-face time with a teacher or tutor in a small group, and some time learning with and from peers. Blended learning often benefits from a reconfiguration of the physical learning space to facilitate learning activities, providing a variety of technology-enabled learning zones optimized for collaboration, informal learning, and individual-focused study.

Cognitive Development

The process of learning, memory, attention, concentration, and language development.

Cohesiveness

The degree to which group members enjoy collaborating with the other members of the group and are motivated to stay in the group.

Conformity

Adjusting one's behavior to align with the norms of a particular group

Control of Institutions

A classification of institutions of elementary/secondary or postsecondary education by whether the institution is operated by publicly elected or appointed officials and derives its primary support from public funds (public control) or is operated by privately elected or appointed officials and derives its major source of funds from private sources (private control).

Counterfactual

Contrary to facts or thinking about alternative possibilities for past or
future events: what might happen/have happened if...?

Critical Thinking

The act or practice of thinking by applying reason and questioning
assumptions to solve problems and evaluate information without the influence of biases, personal feelings, and opinions.

Crosscutting Concepts

Ideas that hold true across many and often disparate disciplines.

Cyberbullying

Bullying is unwanted, aggressive behavior that involves a real or
perceived power imbalance. This power imbalance can be
physical. It can also revolve around popularity or the bully
having access to embarrassing information about the victim.
Generally, bullying is a repeated behavior, or it has the potential to be repeated. Cyberbullying, then, is when these bullying behaviors occur online, either through messaging, social
media, or other digital channels.

Cyberstalking

This highly intimidating form of cyberbullying is when a person
tracks another in the digital sphere and sends them negative
comments, which can include threats, to frighten and terror-
ize them. In some cases, this can even lead to physical stalking
in real life.

Data Point

A discrete unit of information.

Dataset

Related data points in a collection, usually with tags (labels) and a
uniform order.

Digital Citizen

Refers to an individual who engages responsibly, ethically, and positively in the digital world. Being a digital citizen involves
using technology, especially the Internet, in a manner that
respects the rights and well-being of others, while also contributing to the overall betterment of the online community.
Digital citizenship encompasses a range of skills, knowledge,
and attitudes that empower individuals to navigate the digital

landscape safely and effectively. This includes understanding and practicing concepts such as online etiquette, responsible use of information, digital literacy, and awareness of potential risks and challenges in the digital environment.

Digital Citizenship

The safe, ethical, responsible, and informed use of technology. This concept encompasses a range of skills and literacies that can include Internet safety, privacy and security, cyberbullying, online reputation management, communication skills, information literacy, and creative credit and copyright.

Digital Citizenship (2)

The ability to use digital technology and media in safe, responsible, and ethical ways. Digital Citizenship is a set of fundamental digital life skills that everyone needs to have.

Digital Competitiveness

The ability to solve global challenges, innovate, and create new opportunities in the digital economy by driving entrepreneurship, jobs, growth, and impact.

Digital Creativity

The ability to become a part of the digital ecosystem, and to create new knowledge, technologies, and content to turn ideas into reality.

Digital Intelligence (DQ)

A comprehensive set of technical, cognitive, meta-cognitive, and socio-emotional competencies grounded in universal moral values that enable individuals to face the challenges of digital life and adapt to its demands. Thus, individuals equipped with DQ become wise, competent, and future-ready digital citizens who successfully use, control, and create technology to enhance humanity.

Digital Use Divide

Traditionally, the digital divide referred to the gap between students who had access to the Internet and devices at school and home and those who did not. Significant progress is being made to increase Internet access in schools, libraries, and homes across the country. However, a digital use divide separates many students who use technology in ways that transform their learning from those who use the tools to complete the same activities but now with an electronic device (e.g., digital worksheets, and online multiple-choice tests). The digital use divide is present in both formal and informal learning settings and across high- and low-poverty schools and communities.

Digital Natives

Young people who have been born into a virtual reality, view the world differently, and have a "digital footprint," process infographics speedily, but lack basic capacity for interpersonal interactions.

Disabilities, Children with

Those children evaluated as having any of the following impairments and who, by reason thereof, receive special education and/ or related services under the Individuals with Disabilities Education Act (IDEA) according to an Individualized Education Program (IEP), Individualized Family Service Plan (IFSP), or a services plan. There are local variations in the determination of disability conditions, and not all states use all reporting categories.

Autism

Having a developmental disability significantly affects verbal and non-verbal communication and social interaction, generally evident before age 3, that adversely affects educational performance. Other characteristics often associated with autism are engagement in repetitive activities and stereotyped movements, resistance to environmental change or change in daily routines, and unusual responses to sensory experiences. A child is not considered autistic if the child's educational performance is adversely affected primarily because of an emotional disturbance.

Deaf-Blindness

Having concomitant hearing and visual impairments that cause such severe communication and other developmental and educational problems that the student cannot be accommodated in special education programs solely for deaf or blind students.

Developmental Delay

Having developmental delays, as defined at the state level, and as measured by appropriate diagnostic instruments and procedures in one or more of the following cognitive areas: physical development, cognitive development, communication development, social or emotional development, or adaptive development. Applies only to 3- through 9-year-old children.

Emotional Disturbance

Exhibiting one or more of the following characteristics over a long period, to a marked degree, and adversely affecting educational performance: an inability to learn that cannot be explained by intellectual, sensory, or health factors; an inability to build or maintain satisfactory interpersonal relationships with peers and teachers; inappropriate types of behavior or feelings under normal circumstances; a general pervasive mood of unhappiness or depression; or a tendency to develop physical symptoms or fears associated with personal or school problems. This term does not include socially maladjusted children, unless they also display one or more of the listed characteristics.

Hearing Impairment

Having a hearing impairment, whether permanent or fluctuating, which adversely affects the student's educational performance, is not included under the definition of "deaf" in this section.

Intellectual Disability

Having significantly subaverage general intellectual functioning, existing concurrently with defects in adaptive behavior and manifested during the developmental period, which adversely affects the child's educational performance.

Multiple Disabilities

Having concomitant impairments (such as intellectually disabled-blind, intellectually disabled-orthopedically impaired, etc.), the combination of which causes such severe educational problems that the student cannot be accommodated in special education programs solely for one of the impairments. The term does not include deaf-blind students.

Orthopedic Impairment

Having a severe orthopedic impairment that adversely affects a student's educational performance. The term includes impairment resulting from congenital anomaly, disease, or other causes.

Other Health Impairments

Having limited strength, vitality, or alertness due to chronic or acute health problems, such as a heart condition, tuberculosis, rheumatic fever, nephritis, asthma, sickle cell anemia, hemophilia, epilepsy, lead poisoning, leukemia, or diabetes, which adversely affect the student's educational performance.

Specific Learning Disability

Having a disorder in one or more of the basic psychological processes involved in understanding or in using spoken or written language, may manifest itself in an imperfect ability to listen, think, speak, read, write, spell, or do mathematical calculations. The term includes such conditions as perceptual disabilities, brain injury, minimal brain dysfunction, dyslexia, and developmental aphasia. The term does not include children who have learning problems that are primarily the result of visual, hearing, motor, or intellectual disabilities, or of environmental, cultural, or economic disadvantage.

Speech or Language Impairment

Having a communication disorder, such as stuttering, impaired articulation, language impairment, or voice impairment, that adversely affects the student's educational performance.

Traumatic Brain Injury

Having an acquired injury to the brain caused by an external physical force, resulting in total or partial functional disability psychosocial impairment or both, that adversely affects the student's educational performance. The term applies to open or closed head injuries resulting in impairments in one or more areas, such as cognition; language; memory; attention; reasoning; abstract thinking; judgment; problem-solving; sensory, perceptual, and motor abilities; psychosocial behavior; physical functions; information processing; and speech. The term does not apply to brain injuries that are congenital or degenerative or to brain injuries induced by birth trauma.

Visual Impairment

Having a visual impairment that, even with correction, adversely affects the student's educational performance. The term includes partially seeing and blind children.

Disability

Term for impairments, activity limitations, and participation restrictions.

Dissing

When someone spreads false or negative information about a person to damage their reputation.

Doxxing

See Trickery

Elementary School

A school that offers more kindergarten through grade 4 than grades 5 through 8 and no grades 9 through 12.

Emotional Regulation

An individual's ability to manage and respond to emotional experiences such as stress, anxiety, mood, temperament, and hyperactivity/impulsivity.

Empathy

To understand, be aware, and be sensitive to the experiences feelings, and thoughts of another person.

Energetic Play

Active play is equivalent to moderate-to-vigorous physical activity, when children get out of breath and feel warm. This may take many forms and may involve other children, caregivers, objects, or not.

Equity

In education means increasing all student's access to educational opportunities with a focus on closing achievement gaps and removing barriers that students face based on their race, ethnicity, or national origin; sex; sexual orientation or gender identity or expression; disability; English language ability; religion; socioeconomic status; or geographical location.

Flaming

When extreme or offensive language is used to try and cause stress or anxiety to the victim.

Fraping

When someone pretends to be someone else online and posts something silly or inappropriate. It can be intended as a joke but in some cases, it can cause serious negative repercussions.

Gender Identity

One's inner sense of one's own gender, which may or may not match
the sex assigned at birth.

Generative Artificial Intelligence

Technology that creates content — including text, images, video,
and computer code — by identifying patterns in large
quantities of training data, and then creating new, original material that has similar characteristics. Examples
include ChatGPT for text and DALL-E and Midjourney
for images.

Grooming (a.k.a. Cyber Grooming)

The process of "befriending" a young person online to facilitate online
sexual contact and/or a physical meeting with them with the
goal of committing sexual abuse.

Harassment

Constant online abuse can occur on messaging apps or via comments on
social media sites, chat rooms, or gaming sites. Conduct that is
unwelcome and denies or limits a student's ability to participate
in or benefit from a school's education program. The conduct
can be verbal, nonverbal, or physical and can take many forms,
including verbal acts and name-calling, as well as nonverbal
conduct, such as graphic and written statements, or conduct
that is physically threatening, harmful, or humiliating.

Homeschool Students

Students are considered to be homeschooled if all of the following conditions are met: their parents reported them being schooled at
home instead of at a public or private school, their enrollment
in public or private schools did not exceed 24 hours a week
(or 25 hours a week before 2019), and they were not being
homeschooled only due to a temporary illness. Homeschooled
students include those ages 5–17 with a grade equivalent of
kindergarten through grade 12.

Identity Theft

When a cybercriminal steals someone's personal information and uses
it to assume their identity. This can involve the criminal applying for credit and loans, or even filing taxes using the victim's
identity, potentially damaging their credit status.

Individuals with Disabilities Education Act (IDEA)

A federal law was enacted in 1990 and reauthorized in 1997 and 2004. IDEA requires services for children with disabilities throughout the nation. IDEA governs how states and public agencies provide early intervention, special education, and related services to eligible infants, toddlers, children, and youth with disabilities. Infants and toddlers with disabilities (birth–age 2) and their families receive early intervention services under IDEA, Part C. Children and youth (aged 3–21) receive special education and related services under IDEA, Part B.

Impersonation

Creating a fake social media account or email address to pretend to be someone else and then using it for negative purposes.

Intimidation

Placing another person in reasonable fear of bodily harm through the use of threatening words and/or other conduct, but without displaying a weapon or subjecting the victim to actual physical attack.

Machine Learning (ML)

A subcategory of artificial intelligence, in which the study or the application of computer algorithms is designed to improve automatically through experience. Machine learning algorithms build a model based on training data to perform a specific task, like aiding in prediction or decision-making processes, without necessarily being explicitly programmed to do so.

Masquerading

See Impersonation

Mentor

A role model for the child. Offers encouragement and support, and the child trusts the mentor to assist them with difficult situations or concepts.

Metabolic Equivalent of Task (MET)

The metabolic equivalent of task, or simply metabolic equivalent, is a physiological measure expressing the energy cost (or calories) of physical activities. One MET is the energy equivalent expended by an individual while seated at rest.

Middle School

A school that offers more grades 5 through 8 than higher or lower grades but does not offer both kindergarten through grade 4 and grades 9 through 12.

Multi-Tiered System of Supports (MTSS)

A proactive and preventative framework that integrates data and instruction to maximize student achievement and support students' social, emotional, and behavioral needs from a strengths-based perspective.

Neural Network

Also known as an artificial neural network, neural net, or deep neural net; a computer system inspired by living brains.

Neuroscience

Examines the structure and function of the human brain and nervous system. The study of neuroscience in education and learning is rapidly developing.

Non-Screen-Based Sedentary Time

Usually refers to time spent sitting, not using screen-based entertainment. For young children, this includes lying on a mat, sitting in a high chair, pram, or stroller with little movement, sitting reading a book, or playing a sedate game.

Norms

The acceptable standards of behavior within a group, are shared by the members.

Openly Licensed Educational Resources

Teaching, learning, and research resources that reside in the public domain or have been released under a license that permits their use, modification, and sharing with others. Open resources may be full online courses, digital textbooks, or more granular resources such as images, videos, and assessment items.

Organization for Economic Cooperation and Development (OECD)

An intergovernmental organization of industrialized countries that serves as a forum for member countries to cooperate in research and policy development on social and economic topics of common interest. In addition to member countries, partner countries contribute to the OECD's work in a sustained and comprehensive manner.

Outcomes

Something that follows as a result or consequence, a conclusion reached through a process of logical thinking.

Outing

See Trickery

Password

A confidential and alphanumeric sequence of characters that serves as a form of authentication, allowing an individual to access a computer system, online account, or digital device. The user must provide the correct password, which is known only to them, to verify their identity and gain authorized entry. Passwords play a crucial role in securing digital information and preventing unauthorized access to personal or sensitive data.

Password Hygiene

Creating unique and separate passwords for sensitive online accounts, managing passwords using browser or stand-alone applications, and the tactics of changing passwords.

Passphrase

A sequence of words or other text used as a form of authentication to access a computer system, online account, or digital device. Unlike traditional passwords, passphrases are typically longer and consist of multiple words or a combination of words, numbers, and symbols. Passphrases enhance security by creating a more complex authentication mechanism, making it harder for unauthorized individuals or automated programs to guess or crack the access credentials.

Personalized Learning

Refers to instruction in which the pace of learning and the instructional approach are optimized for the needs of each learner. Learning objectives, instructional approaches, and instructional content (and its sequencing) may all vary based on learner needs. In addition, learning activities are meaningful and relevant to learners, driven by their interests, and often self-initiated.

Pervasive Computing

Also known as ubiquitous computing, refers to the integration of computing technology into everyday objects and environments, making them smart, connected, and capable of interacting with each other and with users seamlessly. The goal of pervasive computing is to create an environment where computational power is embedded in the fabric of daily life, making it unobtrusive and enhancing user experiences.

Pervasive Education

Typically refers to the integration and widespread presence of educational technologies, digital tools, and learning opportunities throughout various aspects of a learner's life and environment. It encompasses the idea that education is not confined to traditional classroom settings but extends into the broader context of a learner's daily experiences.

Pervasive Technology

Refers to the widespread integration and seamless presence of technology in various aspects of daily life and across diverse environments. In a pervasive technology landscape, digital tools, devices, and systems become an integral and often unnoticed part of the fabric of society, enhancing and influencing the way people live, work, and interact.

Positive Behavioral Interventions and Supports (PBIS)

A framework for creating safe, positive, equitable schools, where every student can feel valued, connected to the school community, and supported by caring adults.

Prekindergarten

Preprimary education for children typically ages 3–4 who have not yet entered kindergarten. It may offer a program of general education or special education and may be part of a collaborative effort with Head Start.

Preschool

An instructional program enrolling children generally younger than 5 years of age and organized to provide children with educational experiences under professionally qualified teachers during the year or years immediately preceding kindergarten (or before entry into elementary school when there is no kindergarten).

Project-Based Learning

Takes place in the context of authentic problems, continues over time, and brings in knowledge from many subjects. Project-based learning, if properly implemented and supported, helps students develop 21st-century skills including creativity, collaboration, and leadership, and engages them in complex, real-world challenges that help them meet expectations for critical thinking.

Public Charter School

A school providing free public elementary and/or secondary education to eligible students under a specific charter granted by the state legislature or other authority and designated by such authority to be a charter school.

Repurpose

The use of something for a purpose other than its original intended use, or the purpose for which it has been currently used.

Response to Intervention (RTI)

A data-driven approach schools use to support students academically. This approach is characterized by analyzing assessment data, planning high-quality instruction, implementing interventions, and evaluating each student's response to interventions.

Role

A set of expected behavior patterns attributed to someone occupying a given position in a social unit.

Secondary/High School

A school that offers more grades 9 through 12 than grades 5 through 8 and no kindergarten through grade 4.

Sedentary Screen Time

Time spent passively watching screen-based entertainment (TV, computer, mobile devices). Does not include active screen-based games where physical activity or movement is required.

Social and Emotional Learning (SEL)

A strengths-based, developmental process that begins at birth and evolves across the lifespan. The process through which children, adolescents, and adults learn skills to support healthy development and relationships.

Science, Technology, Engineering, and Mathematics (STEM)

Fields of study that are considered to be of particular relevance to advanced societies. For The Condition of Education, STEM fields include biological and biomedical sciences, computer and information sciences, engineering and engineering technologies, mathematics and statistics, and physical sciences and science technologies. STEM occupations include computer scientists and mathematicians; engineers and architects; life, physical, and social scientists; medical professionals; and managers of STEM activities.

Tribal Colleges and Universities

An institutional classification developed by the Andrew W. Carnegie Foundation for the Advancement of Teaching. Tribal colleges and universities, with few exceptions, are tribally controlled and located on reservations. They are all members of the American Indian Higher Education Consortium.

Trickery

When someone is groomed to build a relationship of trust and trick them into providing personal details or photos — they are then used to humiliate the person.

Trolling

Making constant comments to try and get a reaction out of the person. Some trolling can be intended as a harmless joke but often it has a malicious intent.

Trustworthy Artificial Intelligence (AI)

AI systems that exhibit characteristics like resilience, integrity, security, and privacy if they're going to be useful and people can adopt them without fear.

References

The terms and definitions identified in this glossary are based on a series of international reference documents and were developed, in part, using selected material from but not limited to, the following resources.

Karppinen, I., Nurse, J.R.C., Varughese, J., 2023, *Oh Behave! The Annual Cybersecurity Attitudes and Behaviors Report 2023*, The National Cybersecurity Alliance and CybSafe, www.cybsafe.com/whitepapers/cybersecurity-attitudes-and-behaviors-report

National Center for Education Statistics (NCES), 2023, *Children's Internet Access at Home. Condition of Education, Glossary*, U.S. Department of Education, Institute of Education Science, https://nces.ed.gov/programs/coe/indicator/cch., and https://nces.ed.gov/programs/coe/glossary#race

(n.a.), 2023, *Essential Components of MTSS*, American Institutes for Research, https://mtss4success.org/essential-components#:~:text=A%20multi%2Dtiered%20system%20of,from%20a%20strengths%2Dbased%20perspective.

(n.a.), 2023, *What is PBIS?*, Center on PBIS, Positive Behavioral Interventions & Supports, www.pbis.org., www.pbis.org/pbis/what-is-pbis.

(n.a.), 2021, *RTI vs. MTSS*, Interval Technology Partners, LLC, www.enriching-students.com/rti-vs-mtss/#:~:text=RTI%20is%20considered%20a%20more,%2C%20and%20social-emotional%20support.

(n.a.), January 21, 2020, *Trustworthy AI: A Q&A with NIST's Chuck Romine*, www.nist.gov/blogs/taking-measure/trustworthy-ai-qa-nists-chuck-romine.

Park, Y., 2019, DQ Global Standards Report 2019, Common Framework for Digital Literacy, Skills, and Readiness, DQ Institute, www.dqinstitute.org/dq-framework.

World Health Organization, 2019, *Guidelines on Physical Activity, Sedentary Behaviour, and Sleep for Children Under 5 Years of Age*, ISBN 978-92-4-155053-6.

(n.a.), January 2017, *Reimagining the Role of Technology in Education: 2017 National Education Technology Plan Update*, U.S. Department of Education, Office of Educational Technology, Washington, DC, Department of Education, http://tech.ed.gov.

(n.a.), (n.d.), *How to Introduce the Kids to the Internet*, Virgin Media, www.virginmedia.com/blog/online-safety/childrens-internet-safety-test.

Index

Note: *Italic* page numbers refer to figures.

Printed in the United States
by Baker & Taylor Publisher Services